The Promise of Nostalgia

The Promise of Nostalgia analyses a range of texts – including *The Virgin Suicides*, both the novel by Jeffrey Eugenides' and Sofia Coppola's screen adaptation, photography of Detroit's 'abandoned spaces', and blogger Tavi Gevinson's media output – to explore nostalgia as a prominent affect in contemporary American cultural production.

Counter to the prevalent caricature of nostalgia as anti-future, the book proposes a more nuanced reading of its stakes and meanings. Instead of understanding it as evidence of the absence of utopia it contends that there is a masked utopian impulse in this nostalgia 'mode' and critical potential in what has typically been dismissed as ideological.

This book will be of interest to scholars, graduate students and upper-level undergraduate students interested in contemporary culture, cultural theory, media studies, the Frankfurt School, utopian studies and American literature and culture.

Nicola Sayers recently completed her PhD at the School of Advanced Study, University of London, UK.

Routledge Research in Anticipation and Futures
Series editors: Johan Siebers and Keri Facer

The Promise of Nostalgia
Reminiscence, Longing and Hope in Contemporary American Culture
Nicola Sayers

www.routledge.com/Routledge-Research-in-Anticipation-and-Futures/book-series/RRAF

The Promise of Nostalgia

Reminiscence, Longing and Hope in Contemporary American Culture

Nicola Sayers

Routledge
Taylor & Francis Group
LONDON AND NEW YORK

earthscan
from Routledge

First published 2020
by Routledge
2 Park Square, Milton Park, Abingdon, Oxon OX14 4RN

and by Routledge
52 Vanderbilt Avenue, New York, NY 10017

Routledge is an imprint of the Taylor & Francis Group, an informa business

British Library Cataloguing-in-Publication Data
A catalogue record for this book is available from the British Library

Library of Congress Cataloging-in-Publication Data
Names: Sayers, Nicola (Author of The promise of nostalgia), author.
Title: The promise of nostalgia : reminiscence, longing and hope in contemporary American culture / Nicola Sayers.
Description: Milton Park, Abingdon, Oxon ; New York : Routledge, 2020. | Series: Routledge research in anticipation and futures | Includes bibliographical references and index. |
Identifiers: LCCN 2019033140 (print) | LCCN 2019033141 (ebook) | ISBN 9780367134983 (hardback) | ISBN 9780367134990 (ebook)
Subjects: LCSH: Nostalgia–United States. | Nostalgia in mass media. | Utopias in mass media.
Classification: LCC BF575.N6 S39 2020 (print) | LCC BF575.N6 (ebook) | DDC 155.9–dc23
LC record available at https://lccn.loc.gov/2019033140
LC ebook record available at https://lccn.loc.gov/2019033141

ISBN: 978-0-367-13498-3 (hbk)
ISBN: 978-0-367-13499-0 (ebk)

Typeset in Bembo
by Wearset Ltd, Boldon, Tyne and Wear

MIX
Paper from responsible sources
FSC
www.fsc.org
FSC® C013056

Printed and bound in Great Britain by
TJ International Ltd, Padstow, Cornwall

Contents

Figures

Preface

Linda Hutcheon, writing about nostalgia in the 1990s, felt the need to confess: 'I am utterly un-nostalgic […] a personality fault to which I must admit.'[1] By contrast, I confess: I am a nostalgic, and I am a daydreamer.

Shortly after my twenty-seventh birthday, in 2007, I moved from London to New York. I knew almost no one, and knew little of the city. Excited and apprehensive in equal measure, I found myself a room in the then still somewhat edgy Williamsburg. By day, I rode the L-train to and from my classes at NYU, clutching worn copies of Sigmund Freud, Paul Tillich and Rollo May. Sitting in my subway seat, sipping my morning latte and peering out over the many-layered scarves I donned in a futile attempt to protect me from the bitter January cold, I suppose that I seemed very much a piece with my environment.

By night, though, I entered a different world. Alone in my room, I lay those first few months and watched – from start to finish – all six seasons, all 128 episodes, of *Dawson's Creek*: a teen drama set in the 1990s in a fictional suburbia, temporally and culturally far from the vibrant metropolitan scene on my doorstep. And as I watched and watched, I was consumed by an indefinable longing that lay just beyond the grasp of clarity – an ache, almost. I was consumed, most of all, by nostalgia. I was nostalgic for the feelings I had had when watching the programme for the first time as a seventeen-year-old: feelings of unknowing and a sense of the possibilities for both romance and study that the beautiful, sincere, studious protagonist Joey Potter had so embodied, whether climbing illicitly into the window of the boy-next-door or staying up late at the library, the autumnal hues of her brown coat and auburn hair flickering in the light. My nostalgia was also for an America that I'd only ever known in the movies, an America that I'd somehow ached for even when I'd first watched *Dawson's Creek*, though I was at the time contemporary both in age and era with the show's characters, and had not then even been to America. It was an America that seemed strangely no closer now that I was an adult living in the City of Dreams than it had been when a schoolgirl in London.

I knew that my nightly pastime was not something to shout about. I had a perception that I ought to have been more critical, and that popular culture was a legitimate object of interest only if it was ironic, or at least consumed somewhat ironically (both *Dawson's Creek* itself and my rabid consumption of it were utterly devoid of irony). I also had a sense, not yet articulated even to myself, that the popular culture that was 'okay to like' was usually male-oriented (comic book adventures were okay, romantic teen dramas were not). I had not yet become interested in theories of nostalgia or utopia, and would not have been able to answer the question of whether nostalgia and utopia were conceptually aligned or opposed. But I did know that both the 'dreamer' and the 'nostalgic' were considered slightly hopeless, figures who were at best indulged and at worst pitied. I wagered, then, that the image of me in this new city, lost in dreams of a 1990s teen suburbia, was best kept to myself.

But I also sensed something else. The ache, the lack, the longing brought about by my self-induced nostalgia festival was at the same time somehow exciting and enlivening. I felt, in reconnecting with my teen fantasies, a strange tension between remembrance and possibility. It was not the fantasies themselves that moved me, but the renewed stirring of an orientation towards the world which I'd had once but had forgotten, or at least muted, during the intervening years of numbing work and dawning adulthood. I was reminded of 'strange twinklings' and 'pregnant echoes' to which I began, slowly, to respond.[2] Although it defied conscious explanation, these embarrassing, trivial, Hollywood-inspired daydreams of my yesteryear were subtly prompting me to respond anew, or differently, to my possible today and tomorrow.

In the years since, I became fascinated by the question of utopia. I read Fredric Jameson, Terry Eagleton, Slavoj Žižek, David Harvey and Susan Sontag, and discovered that utopia was more than the staid thought experiments I had previously all but ignored in Thomas More and others like him: it was an urgent question. I noticed in these thinkers a worry that resonated. If the twentieth century had been characterised by utopian experiments (albeit largely failed experiments), was the twenty-first century characterised by the absence of utopia? Had belief in the possibility of change – belief in the future – all but vanished? Had Margaret Thatcher's maxim that 'there is no alternative' become an unshakable given? In this exploration of utopia, I was led also to Walter Benjamin, Herbert Marcuse and Ernst Bloch: writers so rich that I am still frequently moved – in the original sense of having one's position altered irrecoverably – by the beauty or brilliance of a single sentence.

It was in my quest to understand utopia that I stumbled on nostalgia. If utopia was relatively absent in the twenty-first century (at least in America), nostalgia, once I started to notice it, seemed to be everywhere: on TV and blogs, in fashion, food, film, merchandise and art. More than just a personal predilection, then, nostalgia appeared to be a cultural phenomenon. I started gradually to view my own private longings not merely as a mildly embarrassing quirk, but as connected with 'little daydreams' being harboured in bedrooms across America: millions of tiny lights flickering, singing a tune that, like my *Dawson's Creek* nostalgia, was just slightly off-kilter, not entirely contemporary with the actually lived moment.[3]

If the political Right has given utopia a bad name, dismissing it as escapist nonsense, I began to notice that nostalgia was often given a hard time by the Left. Choose a side, they seem to say: change or the status quo, possibility or regression, the future or the past. But this didn't resonate with my own experience; the nostalgic and the dreamer in me were too closely aligned. Moreover, I thought of those millions of tiny lights flickering, of that hum of nostalgic longing that seemed to permeate American popular culture, and wondered if it was wise to too quickly dismiss these heartfelt stirrings. I also wondered how these little longings connected with the bigger question that consumed me: where had utopia gone? Was it stuck somewhere in the past, where it could be re-lived only in endless media repetitions? And if so, what did that mean for the future?

What follows is my attempt to make sense of some of these questions. It takes me, and the reader, to some strange places: to the Illinois bedroom of a teen blogger named Tavi and to the fiction of five beautiful girls committing suicide in 1970s suburbia; to the abandoned factories of today's Detroit and to the run-down arcades of Walter Benjamin's Paris; to the philosophical musings of Jewish-German intellectuals of the twentieth century and to imagery used in twenty-first century advertising content. But the hidden heartbeat of this inquiry, the pulse that is discernible throughout, is the 'life-light, containing nothing stale' that 'shines vexatiously', 'the freshness, the otherness', the dream 'that does not stop inserting itself into the gaps': hope.[4]

Notes

1 Linda Hutcheon, 'Irony, Nostalgia and the Postmodern', in *Methods for the Study of Literature as Cultural Memory*, ed. by Raymond Vervliet and Annemarie Estor (Boston: Brill, 2000), pp. 189–207 (p. 190).

2 Craig Hammond, 'Towards a Neo-Blochian Theory of Complexity, Hope and Cinematic Utopia' (unpublished doctoral thesis, University of Lancaster, 2012), pp. 15–16.

3 Ernst Bloch, *The Principle of Hope*, trans. By Neville Plaice, Stephen Plaice and Paul Knight, 3 vols (Cambridge, MA: The MIT Press, 1986), I, p. 19.

4 Bloch, *The Principle of Hope*, I, p. 28, p. 22, p. 29.

Acknowledgements

It has been a great privilege to have spent these years immersed in a topic that fuels both my intellectual curiosity and heartfelt engagement, and I am thankful to the many people who supported and inspired me in this pursuit. To my friends (badass women, all): Camilla Buttery, Megan Walsh, Kath Brinson, Claudia Renton, Lucy Davis and Maryam Omidi. To my brothers, Freddie Sayers and Henry Brougham, for humouring me and keeping my feet at least somewhat on the ground. To Tavi Gevinson and Petra Collins, for their generosity in granting permission to use images. To Catherine Davies, for her enthusiasm and counsel. To Johan Siebers, for his insight, intelligence, kindness, and resistance to instrumental reason. To Jarad, whose love, patience and care underwrites every word I ever write. *For* Joseph and Aurora, my lights. And to my parents, for everything.

Introduction

In 1973 the film *American Graffiti* was released. Set in 1962 California, it follows a group of teenagers on their last day of school. Intentionally situated before many of the disruptions of the 1960s – the Vietnam War, the assassination of the Kennedy brothers and Martin Luther King, the ghetto riots, student protests, the Civil Rights movement and more[1] – it harks back nostalgically to a seemingly more innocent era, typified by rock 'n' roll and red Chevrolets. The film set off what one journalist, a teenager at the time of the film's release, describes as a 'good time cultural frenzy': 'for a while, everyone played '50s music, danced '50s dances, and tried to pretend they were living inside *American Graffiti*'.[2]

Twenty years later, in 1993, director Richard Linklater released a film, *Dazed and Confused*, which follows a group of high school students on their last day of school in 1976. Although different in tone, the parallels with *American Graffiti* are clear. This time, however, the film's audience longed for the 1970s, the time when people like them were in reality busy watching *American Graffiti* and longing for the 1950s and early 1960s. But they long in a similar way: the music, clothing, cars and pop-cultural references all conspire to create a mood. Linklater's 'detail of the recreation of the past is extreme', and has its desired effect: the audience is gripped by nostalgia.[3]

In 2007 an eleven-year-old living in suburban Illinois, Tavi Gevinson, started a blog. Within a few years, *Style Rookie* became a huge success and catapulted Tavi into a kind of fame; she was widely celebrated for her intelligent and original voice. Many things could be said about *Style Rookie*, but one of its most striking features is that it is permeated by nostalgia for an amorphous suburban America of an unspecified past. Both in imagery and prose, it presents a vision of dreamy Americana inspired by film and fiction. Notably, the evident nostalgia for this America is divorced from any actual experience of it. It is not personal nostalgia based on memory that this teenager pours out, but nostalgia for an America that belongs to the cultural imaginary.

In 2016 Netflix released an eight-part TV programme, *Stranger Things*, which is described by one reviewer as 'a spooky shot of 80s nostalgia straight to your heart'.[4] Significantly, the nostalgia in this instance is not for the actual 1980s, even as popularly conceived, but for 'the celluloid storytellers who dominated the era'.[5] In *Stranger Things*, the bikes 'are straight out of Spielberg', the romantic moments are 'a touch of Hughes' and 'the rest is Stephen King'.[6] This is nostalgia, then, not for an earlier reality, nor even for a mediated representation of an earlier reality (as in Tavi Gevinson's blog), but nostalgia for earlier media themselves as well as for the cinematic tropes of their moment.

These are all examples of nostalgia culture: cinematic, televisual, digital and literary works that exhibit, feed off and fuel feelings of nostalgia. Of course, there are countless other examples that fall in this bracket, and one can define or limit the contours of the subject matter in different ways, but I begin anecdotally in the belief that the visceral sense of a cultural or literary phenomenon is often best communicated via particular instances.

What is striking, nearly two decades into the twenty-first century – a once much anticipated threshold marking the future – is how much of today's cultural production romanticises the past. In the USA, there is seemingly no end to the ongoing creation of films, TV series, works of literature and other media that look nostalgically to an America of earlier decades. Most often, in the American context, the object of nostalgia is the decade between the 1950s and the 1990s. It might be a specific decade as, for example, in *Dazed and Confused* (1993), *My Girl* (1991), *Summer of Sam* (1999), *The Ice Storm* (1997) and *The Virgin Suicides* (1999), films which look back nostalgically to the 1970s; or it might be a more temporally blurred vision of the American past, as in Tavi Gevinson's blog. Sometimes the nostalgia is for an even earlier era (the TV series *Boardwalk Empire*, for example, evokes the 1920s and 1930s) and sometimes it is for a more recent period (a 2013 call in *Rookiemag*, the online magazine started by Tavi Gevinson, asks for submissions which evoke 'the vintage, old-timey feel of static TV from 2002').[7] Occasionally, this kind of nostalgia is even evident in media or technologies whose content is the present day. For example, Hipstamatic, a digital photography application released in January 2012, allows users of Apple's digital iPhone camera to filter their digital photographs in order to create the effect of their having been taken with a vintage film camera, thus conveying an aura of pastness and prompting the viewer to feel nostalgic about images of the present.

Everybody's *still* just wild about nostalgia

I am not the first person to observe that nostalgia culture is widespread in America, and has been for some time – far from it. Since the 1970s, when an initial 'nostalgia boom' was first widely remarked upon, journalists and scholars alike have been busy noticing and analysing this phenomenon, some even going so far as to diagnose America as having a 'nostalgic condition'.[8]

In February 1971 the cover of *Life* magazine announced 'Everybody's Just Wild About Nostalgia'.[9] The issue is full of the kinds of articles that appeared frequently in the 1970s: enjoyable retrospectives of the recent cultural past, alongside claims that 'Old is in'.[10] A 1972 *Newsweek* cover featured Marilyn Monroe and reported on the 'yearning for the Fifties, the good old days'.[11]

More soberly, a 1970 *New York Times* article discussed the new industry created by nostalgia for extinct pop culture, listing various companies that had flourished in this new environment (Radio Yesteryear, Nostalgia Press, 'Good Old Days' magazine).[12] Another article comments that the nostalgia boom was turning junk into 'junque': 'pieces of salvage that a few years ago would have been prime candidates for a junkman's cart [...] have quietly become big business'.[13]

A few years later, as is typical in relation to popular trends, academic articles on the same topic started appearing. A 1977 article by Fred Davis in *The Journal of Popular Culture* analyses the 'nostalgia wave' of the 1970s, explaining it as a response to the massive identity dislocation brought about by the great upheavals of the 1960s.[14] At the end of the decade, in 1981, Fredric Jameson's now famous critique of the 'nostalgia mode' was first published.[15] As Sprengler points out, 'the question "Why is America so nostalgic?" posed only occasionally earlier in the century, became a mantra in the 1970s'.[16]

But it was not a mantra that ended in the 1970s. The 1970 *New York Times* article on the nostalgia industry ends with the following reflection:

> Perhaps at the end of the twentieth century, the personalities and paraphernalia of today will become the nostalgic items of that time. Can you imagine a sentimental look back at the Beatles, the mini-skirt, Barbra Streisand and the gasoline engine?[17]

The journalist seems incredulous. This is no doubt in part an existential amusement at the notion that what is current will become old, though it also seems to express disbelief that the then prominent nostalgia fad would continue for several decades. But continue it did. The article reads amusingly today because (with the exception of the gasoline engine) the answer to the question posed by the journalist is, of course, yes. The Beatles, mini-skirts and Barbra Streisand are all now subject to sentimental nostalgic remembrance, so much so that the picture unwittingly painted is almost a pastiche of what we now fondly recall as 'the Sixties'.

By the mid-1980s, there was already a journalistic backlash against the so-called nostalgia wave. Michael Kammen examines newspapers of the time and points out that around 1985 'warnings started to flash' about nostalgia.[18] For example, he mentions a 1989 *New York Times* article about the National Association for the Advancement of Time (NAFTAT), an organisation started by three young men in their twenties with a simple stated goal: 'Let's make nostalgia a thing of the past.' NAFTAT puts forward the argument that 1980s pop culture 'is so firmly entrenched in the '60s that kids who would otherwise be discovering the world around them are wishing they had gone to Woodstock'.[19]

By the late 1990s the continuing pervasiveness of nostalgia (and ennui with said nostalgia) is evidenced in the satirical online magazine *The Onion*'s headline 'U.S. Dept. of Retro Warns: "We May Be Running Out of Past"'.[20] Of course, the past has not run out but, more surprisingly, neither has the fascination with the past; and as the production of nostalgia culture has continued, so has the question: Why is America so nostalgic? Hutcheon, for example, wondered in the 1990s about the relationship of nostalgia to the upcoming millennium, but even the dawn of a new century did not put an end to the allure of nostalgia.[21] Sprengler observes that 'the nostalgia thought to have been activated by fin-de-millennium anxieties certainly has not subsided'.[22] As Boym argues:

> The first decade of the twenty-first century is not characterized by the search for newness, but by the proliferation of nostalgias that are often at odds with one another. Nostalgic cyberpunks and nostalgic hippies, nostalgic nationalists and nostalgic cosmopolitans, nostalgic environmentalists and nostalgic metrophiliacs (city lovers) exchange pixel fire in the blogosphere.[23]

Even now, in the second decade of the twenty-first century nostalgia culture is still going strong. For example, a 2015 article in the *National Post* about the long-running TV series *Mad Men* describes the show as 'one long meditation on the central obsession of our time: nostalgia', and characterises nostalgia as 'the opiate of choice for the digital masses'.[24] Nostalgia culture is still so widespread that a 2012 *Time* magazine article describes ours as:

> A moment consisting of some modern elements and samples of previous eras and their aesthetics, like a moment playing dress-up in someone else's clothes rather than expressing its own style.[25]

If America can be diagnosed as having a 'nostalgic condition', as Grainge suggests, then it is one that shows no sign of abating.[26] Moreover, commentary on the cultural phenomenon keeps pace with the phenomenon itself. Lowenthal, already in 1989, wondered if 'perhaps our epoch is awash not in nostalgia but in a widespread *preoccupation* with nostalgia amongst intellectuals and the mass media'.[27] Although the claim is tenuous given the evident expansion in nostalgic cultural production since the 1970s, Lowenthal is right to highlight that the interest in nostalgia culture is part of the nostalgia trend, not independent of it.

Despite the long-standing and ongoing observations of a nostalgia boom in American culture, there has been a consistent tendency to perceive this boom as something new. Niemeyer, for example, in the introduction to the 2014 collection of essays *Media and Nostalgia: Yearning for the Past, Present and Future*, writes:

> Who would have thought, given the 1990s' imagining of a future filled with technology, that the beginning of the new century would in fact be

> marked by an increase in expressions of nostalgia, and in nostalgic objects, media content and styles? This volume goes back to a simple observation of the current nostalgia boom.[28]

There is little recognition here that, far from being filled only with future-oriented imaginings, the 1990s were already home to a substantive nostalgia culture. Indeed, in the 1990s, Hutcheon contrasted the 'nostalgic' 1990s with the 'ironic' 1980s, forgetting that, as Kammen points out, 1985 was a peak year for media commentary on the nostalgia wave.[29]

Furthermore, there is often disagreement in the media debate about what era is the primary nostalgic object, with journalists sometimes mistaking personal preference for general trends. One *Huffington Post* blogger, for example, argued in 2013 that *Dazed and Confused* would not have worked today because the difference between the 1970s (when *Dazed and Confused* was set) and the 1990s (when *Dazed and Confused* was made) was sufficient to warrant nostalgic reminiscence, whereas the difference between 2013 and the 1990s is, in his view, less significant. He therefore doubts whether there is 'anyone feeling nostalgic for 1996'.[30] In fact, there are many people feeling exactly this. As a 2014 article in *I-D* magazine puts it, 'Scroll through the feeds of fashion bloggers, stylists, designers and everyone who follows them and all you'll find is an endless stream of Kate Moss #TBTs [...] perhaps we are the least optimistic and most sentimental generation to have ever existed. Constantly reminiscing [...] about the 90s and the heyday of "Cool Britannia".'[31]

Despite these quibbles and inconsistencies, what has remained consistent since at least the early 1970s is both the existence of a widespread nostalgia culture and an interest in and commentary on the phenomenon both in the media and in academia.[32]

The principal aim of this book is to explore the specific meaning and character of this kind of nostalgia in contemporary American culture. But why is this a topic worth pursuing, given that a fair amount has already been said on the subject? The broadest, and simplest, answer to this question is that I believe these forms of nostalgia have hitherto been incorrectly, or at least insufficiently, understood. But the underlying conviction behind this answer is that it matters how we understand and respond to the longings that inform and underlie popular cultural production and reception; that these longings are often clues to our current systemic condition and to the stakes at play; that we would do well to listen, and listen carefully, to what philosopher Ernst Bloch calls the 'little daydreams' that may just be indicators of more all-encompassing realities.

Nostalgia versus utopia

Svetlana Boym, a well-known recent theorist of nostalgia, argues that: 'The twentieth century began with a futuristic utopia and ended with nostalgia. Optimistic belief in the future was discarded like an outmoded spaceship sometime in the 1960s.'[33] This echoes Susan Sontag's reflection, in 1996, on the publication of a number of her own essays in *Against Interpretation* (1961) that 'perhaps the most interesting characteristic of the time now labelled the Sixties was that there was so little nostalgia. In that sense, it was indeed a utopian moment. The world in which these essays were written no longer exists'.[34] Aside from Boym's and Sontag's shared characterisation of the contemporary as more nostalgic than utopian, there is another more subtle shared understanding in both of their statements: that nostalgia and utopia are inherently opposed, even mutually exclusive – where one is, the other is not.

This intuitive tension between nostalgia and utopia arises out of a deep-rooted sense that utopia's proper orientation is towards the future, whereas nostalgia is stuck in the past. The historical and semantic backgrounds of this perceived tension will be explored fully in Chapters 1 and 2, but it is immediately evident that if nostalgia is opposed to utopia then nostalgia must be poorly regarded, at least among those who value utopian thinking. And indeed, nostalgia culture has been viewed as highly suspect by most left-leaning theorists, who often associate it with conservatism, ideology and the status quo (or, worse still, regression), whilst associating utopia with hope, the future, and the possibility of social change.

Foremost among these theorists is Fredric Jameson, a committed utopian thinker – his book, *Archaeologies of the Future: The Desire Called Utopia and Other Science Fictions* is dedicated to 'my comrades in the party of Utopia'[35] – whose famous critique of what he calls the 'nostalgia mode' is aligned with the long-standing presumption of an opposition between nostalgia and utopia, which he brings to bear within the context of a depthless postmodern culture. But there are many others who fear that the future is at stake and have a sense that the re-hashing of, and fascination with, outmoded culture is a sign of hopelessness: of an inability, today, to imagine alternative futures. Mark Fisher, for example, fears that 'the young are no longer capable of producing surprises'.[36] (Chapters 1–3 will elaborate the theoretical lay of the land in depth; at this stage it suffices to observe that nostalgia has not typically been positively viewed among advocates of utopian thought as essential to social change.)

The assumed opposition between nostalgia and utopia as that between the past and the future is also evident in popular discourse on the subject. For example, already in the 1980s, the young men who founded NAFTAT did so in resistance to the 'media nostalgia complex', which they feared was crushing contemporary creativity.[37] Similarly, the 2012 *Time* magazine article begins: 'The past is the present. It hangs in the cultural air all around us like a fog clouding our ability to build a unique now.'[38] Whether nostalgia culture is a result of an inability to think the future in contemporary society, or whether it somehow prevents present or future-oriented cultural practice (the new, or the now, in other words) from emerging, nostalgia and utopia are widely presumed to be incompatible.

The intellectual intuition which led me to write this book was that this commonly held view – that, in layman's terms, one cannot simultaneously long back in time *and* believe in or have hope for a better future – is overly simplistic. It seemed to me that this highly prevalent (if at times merely implicit) perspective failed to take into account the complexity of longing, and the nuanced, often non-linear, ways in which past, present and future inform one another. The hard graft of examining this intuition, and of arguing my position as it emerged more clearly during the course of my research, is of course the work of the book as a whole. But it is worth stating up front that my endeavours led me to the German philosopher Ernst Bloch, whose theoretical insights, I have come to believe, are invaluable to the current debate about nostalgia culture as they are, indeed, to contemporary literary and cultural studies more broadly.

A Blochian approach

The Institute of Social Research, also known as the Frankfurt School, was founded in 1923 in Frankfurt, moved briefly to Geneva in 1933 in response to the growing threat of Nazism, and then on to New York, where it became affiliated with Columbia University until the early 1950s. It was then formally re-established in Frankfurt, although many of its leading thinkers chose to stay in the United States. Critical of both Soviet Socialism and Western Capitalism (albeit to varying degrees) – and of course of Nazism – this group of highly original and mostly Jewish-German thinkers was motivated to 'keep alive a space of emancipation during a period that did not offer it'.[39] This proved to be a fertile moment for theorising the utopian impulse. The engagement of these critical theorists with utopia is more than a limited interest in a literary genre or even in explicitly utopian communities; it relates to the broader question of hope, considered from political, philosophical and personal perspectives. Since the Institute was explicitly interdisciplinary, the varied and complex positions of different Frankfurt School theorists vis-à-vis utopia draw on and incorporate a range of traditions of thought – most prominently, but not exclusively, Marxism, German Idealism and Psychoanalysis – which are brought to life in new ways and new contexts.

Significantly for our purposes, several of these theorists offer complex meditations on the relationship between memory, hope, nostalgia and utopia, which upset any too-neat dichotomies. Most of them rejected straightforward notions of teleology and progress, and accordingly any overtly futuristic utopian ideals; but they were equally wary of backward-longing romanticism, not least because of the uses they saw Nazism making of these sentiments. Janet Stewart argues that 'they did not so much suspend the utopian motif as search for different methods of approaching utopia.'[40] Stewart articulates three important questions that plagued the critical theorists of the Frankfurt School: If Utopia is not simply located in the future, then where is it and how does one attain it? What is the relationship of the past to the future? How can one articulate utopia if the idea of teleological progress has been rejected?[41] She also argues that the effort of these theorists to (re)discover utopia is bound up with the role of memory, although none of them believes the relationship between memory and utopia is straightforward. They teeter, to varying degrees, between nostalgic remembrance of a lost ideal and suspicion of that remembrance.

There are three thinkers on whom I will draw substantially to elaborate my argument: primarily, Ernst Bloch and, secondarily, Walter Benjamin and Herbert Marcuse. Since my approach relies principally on the work of Bloch I call it 'Blochian' for the sake of ease, but it also incorporates ideas of both Benjamin and Marcuse. Of the three, only Marcuse was a formal member of the Institute of Social Research, but both Bloch and Benjamin are often placed beneath the 'Frankfurt School' banner because of their strong associations with members of the School, and because of an intersection of concerns and shared intellectual work.

Interestingly, despite fascinating overlaps in both their life histories and intellectual concerns, Bloch and Marcuse had little known communication. Their lives spanned a similar period (Bloch: 1885–1977; Marcuse: 1898–1979), both were German Jews who immigrated to America during the Nazi years (Marcuse in 1933 and Bloch in 1938; although, unlike Marcuse, who remained in the United States, Bloch returned to the GDR in 1949), and they shared a common set of friends and intellectual influences. Strangely, however, Marcuse seldom references Bloch's work – although as Levitas points out he could not have been unaware of it.[42]

In contrast, Bloch and Benjamin were friends and interlocutors, often engaging with one another's work, right up until Benjamin's untimely death in 1940. What is more, among their close contemporaries, Bloch and Benjamin are unique in having both combined a thorough commitment to Marxism with a deep interest in mysticism. In their Marxism, both were influenced by their respective close friendships with the playwright Bertolt Brecht. In their mysticism, they differed in influence and orientation. Benjamin, described by Leon Wieseltier as 'modernity's kabbalist' – 'in his turgidly enchanted world there were only mysteries, locked and unlocked'[43] – centred his mysticism on Jewish sacred texts and traditions, and was heavily influenced by his profound and formative friendship with Kabbalist scholar Gershom Scholem. Bloch, by contrast, favoured Christian mysticism, and *The Principle of Hope* is littered with New Testament interpretations, much to the irritation of both Benjamin and Scholem.[44] Despite this difference, however, the effort to infuse Marxism with a kind of mysticism (and to rescue it from what both deemed a prevalent strain of instrumentalised Marxism) is an intriguing commonality at a time when Marxism was explicitly and overtly hostile to any perceived religious doctrine.

Marcuse is in some ways the poster-boy of the 1960s, having during his time enjoyed a popular reception in America as part of the countercultural movement of that era. In light of Sontag's supposition that the 1960s was notable for its lack of nostalgia, it is perhaps surprising that this utopian philosopher of that utopian time is so drawn to memory – something that will be explored in Chapter 2.

Benjamin's fame came only posthumously, particularly in the 1970s and thereafter. Leon Wieseltier writes that, after the publication of *Illuminations* in 1968 and *Reflections* in 1970, 'in the bookshops around Columbia in its roiled years, before Broadway became a boulevard of theory, they were snatched up immediately and read with a hushed fascination'.[45] It is often pointed out that Benjamin speaks somehow to our current digital age, which explains in part his posthumous and delayed reception. Recent theorists of nostalgia, most prominently Svetlana Boym, have also made use of Benjamin's theories.

By contrast, until recently, Bloch was considered too old-fashioned to have any relevance for contemporary literary or cultural studies. David Kaufman argues that while the ongoing relevance of Benjamin has been persistently clear – his work having been constantly renewed, reclaimed and repurposed in the years since his death – Bloch's work appears rather 'to have been fired by the peculiar intellectual passions of the first decades of this century and presents itself as a somewhat antiquarian curiosity'.[46] This view appears in several introductory passages on Bloch. For example, Siebers refers to 'the metaphysical and cultural discontemporaneity of his thinking' and Geohegan notes that 'Bloch's vast system will strike many as positively antediluvian'.[47] This lack of relevance is explained partly as a function of obscure, outdated or politically troublesome content ('there is perhaps too much Schelling in *The Principle of Hope*, and far too much Stalin') and partly as a consequence of his challenging style ('aphoristic simplicity interlaced with a sometimes baroque and completely unabashed complexity; prose poetry combined with what at times seems a turgid verbosity').[48]

However, in spite of his complexity and obscurity there has been, very recently, a renewed interest in Bloch's philosophy of hope. This is evident in the foundation of the Centre for Ernst Bloch Studies at the University of Sheffield in 2009, international conferences such as the Ernst Bloch London Symposium (December 2013) and the fairly recent publication of *The Privatization of Hope: Ernst Bloch and the Future of Utopia* (2013), edited by Peter Thompson and Slavoj Žižek. There has also been an increasing sense (although still on a comparatively small scale) that Bloch, like Benjamin, speaks to our contemporary moment. Žižek says in his preface to *The Privatization of Hope* that Bloch 'is one of the rare figures about whom we can say: fundamentally, with regard to what really matters, he was right, he remains our contemporary, and maybe he even belongs more to our own time than to his own'.[49] In addition, a number of literary and cultural critics have recently put Bloch's ideas to use in a contemporary context.[50]

The three main theorists whose works inform my own argument were writing in a very different intellectual milieu and speaking to a very different set of societal concerns to those that preoccupy theorists today. Whilst they and their Frankfurt School associates tried to comprehend and analyse the triad of Nazism, Stalinism and American-led global capitalism, contemporary theorists are confronted only by the latter, which anyway survives in a quite different form. It might therefore seem odd to believe that these critical theorists speak to contemporary concerns, and perhaps bizarre to pair the dense, obscure philosophy of Ernst Bloch with the dreamy, teenage, suburban nostalgia of teen blogger Tavi Gevinson, as I do in Chapter 4.

However, the in-depth theorisations of memory and utopia of Bloch, Benjamin and Marcuse are remarkable, and have no real equal today; and there are several reasons for regarding their work as particularly helpful for present purposes. First, they believed that the complexities of culture and society could only be understood by drawing on several different disciplines – including psychoanalysis, philosophy, and social theory – an approach which seems especially well suited to certain contemporary mixed-media texts, such as blogs. Second, both Bloch and Benjamin were committed to the notion that the broader truths of a given moment could be read in even the seemingly insignificant details of life and popular culture, a foundational premise of my own explorations. Third, both Bloch and Marcuse were living in America during the post-war years, at a time when many of the building blocks of today's culture – mass advertising, television, Hollywood – were first becoming pervasive. With their outsider eyes they were able to offer astute insights into the emergence of cultural norms and forms which continue to shape American (and global) culture to this day. Finally, several of today's most prominent theorists of contemporary nostalgia culture – Jameson and Boym, for example – are strongly influenced by these earlier German theorists. I will show that despite this influence (or perhaps because of it, since the language and theoretical lineage is similar), the earlier generation can be used to argue against, or build on, the theories of their intellectual descendants.

If this is not clear already, it is worth stating at the outset that my allegiance is broadly to these Frankfurt School thinkers and the Neo-Marxist orientation of their work. This statement is necessary because there are several ideas that they have argued in depth – for example, the widespread dominance of instrumental reason in late capitalist culture and the problems that follow from this – which I do not wish to argue from scratch, but which inform my own perspective. I assume a certain degree of background knowledge in order to explore in detail the specific contribution of these thinkers to theories of utopia and nostalgia; and in order to draw on them in my own examination of contemporary American nostalgia culture.

What is the contemporary?

What do I mean in describing my object of study as *contemporary* American nostalgia culture? Pedro Erber points out that 'the contemporary' is much discussed in recent literature and yet still eludes an agreed-upon definition.[51] He goes on to outline three prominent definitions, although these by no means cover all of the term's possible meanings:

> On the one hand, we have grown accustomed to defining the contemporary as the historical period that succeeds the modern period and as the epoch

> that we happen to inhabit in the early twenty-first century. On the other hand, contemporaneity refers to a relationship between two events, persons, phenomena, etc. that are contemporaneous with each other; that is, they share the same time. In addition, contemporary refers to that which is contemporary with us – whoever and whenever we are – and is in this sense synonymous with the present.[52]

My most straightforward intention in labelling nostalgia culture 'contemporary' aligns with the first of these meanings: I wish to define my period of interest and to acknowledge that the nostalgia texts that I consider are all twenty-first-century texts. Only in Chapter 5 do I consider a slightly earlier text (the chapter deals with not only Sofia Coppola's 1999 film *The Virgin Suicides* but also the 1993 Jeffrey Eugenides novel from which it is adapted). However, since one of the aspects of nostalgia texts that I consider is reception (as opposed, merely, to production), and since the novel, alongside the film, has in recent years enjoyed renewed popularity – watched and read, for example, by a virtual community of teenage girls, who appreciate it self-consciously for the nostalgia it evokes – both warrant inclusion as texts of contemporary nostalgia culture.

However, the term *contemporary* is far from a neutral descriptor, and the question of what is or is not contemporary is therefore one of the concerns of this book. At a first glance, the very phrase contemporary nostalgia culture – while logically coherent – has a strange ring to it. If nostalgia is a longing away from the contemporary, or at least from the present, it is perhaps paradoxical, or at least odd, to label as contemporary those very elements of the culture which express a desire not to be contemporary.

The use of the term *contemporary* is also complicated with regard to the theorists on whom I am drawing in the analysis of this twenty-first-century culture. As already mentioned, Žižek describes Bloch as 'our contemporary', someone who 'belongs more to our own time than his own', whereas Kaufman describes him as an 'antiquarian curiosity'.[53] It is interesting, as an aside, that Bloch, as someone who thought extensively about the role of the non-contemporaneous (his theory of which will be important to my argument), should himself become the subject of discussion regarding whether his work should be perceived as contemporary or non-contemporary. But who is right, and what does Žižek mean in describing Bloch – with whom we do not 'share the same time' – as our contemporary?

A helpful account of the contemporary is offered in Georgio Agamben's celebrated essay 'What is the contemporary?'. Agamben draws on Roland Barthes's comment that 'the contemporary is the untimely' to give a somewhat contrarian account of the contemporary as that which is out of joint with the present. 'Those who are truly contemporary, who truly belong to their time,' he argues, 'are those who neither perfectly coincide with it nor adjust themselves to its demands.'[54] This anachronism, he suggests, allows them to see their own time more clearly than people who are too aligned with the predominant culture of their time.[55] I want to suggest that Žižek's claim that Bloch is our contemporary relates to this understanding of the term: Bloch is someone who can help us to see the present more clearly, by helping us, as Agamben puts it, to dip our pens 'in the obscurity of the present'.[56]

As will become clearer through the development of ideas and arguments in this book, nostalgia culture, which is in some ways non-contemporary, is also, in other ways, wholly contemporary – at least if we understand *contemporary* as Agamben does. Nostalgia culture expresses longings that are at odds with the predominant culture of the present day, and – although some of these longings are for nothing other than false utopias, for what Bloch calls 'booty for swindlers'[57] – others arise out of a genuine lack and reflect ongoing hopes that are muffled but not entirely lost. As such, they help us to see the present more clearly. This all requires explanation in the light of both Bloch's philosophy and the texts under consideration here, and I do not wish to get too far ahead of myself; but it is worth mentioning that the following passage of Agamben's could almost have been written by Bloch, and intimates the critical approach to nostalgia culture that this book will take:

> this urgency is the untimeliness, the anachronism that permits us to grasp our time in the form of a 'too soon' that is also a 'too late'; of an 'already' that is also a 'not yet'. Moreover, it allows us to recognise in the obscurity of the present the light that, without ever being able to reach us, is perpetually voyaging towards us.[58]

Overview

This book is divided into two parts. Part I is an in-depth theoretical analysis of the relationship between nostalgia, utopia, memory and hope. In Part II, I will offer critical readings of several contemporary American nostalgia texts. Part I is necessary in order to explain fully the approach I take to the texts in Part II. It is necessary also, in order to ground my aims vis-à-vis these contemporary texts in a history both of nostalgia culture and of the theoretical deployment of utopia in literary and cultural criticism, and to situate my own efforts and the originality of my contribution accordingly. In current scholarship, there is often a distinction between theoretical work on Bloch (and indeed Benjamin and Marcuse) and attempts to draw on his theories in undertaking cultural or literary criticism. The latter efforts usually present only a brief introduction to theoretical ideas that they use before beginning their own critical work. I want to suggest that the analyses in Part I and in Part II of this book are complementary and intertwined: a thorough theoretical grounding makes clear how I will approach these contemporary texts, but only particular readings of actual instances of nostalgia culture can reveal anything about its truth in this moment.

Chapter 1 presents an overview of existing theories of, and ideas about, nostalgia, as well as a history of the term's use. It asks: How did nostalgia, originally a term for a specific medical condition, come to be understood as a symptom of the modern age? Why has nostalgia alternatively been described as an anti-modern sentiment and a peculiarly modern phenomenon? And what is meant by the recent postmodern understanding of nostalgia as a consumer mode? The chapter also raises difficulties with, and outstanding questions for, existing theorisations of nostalgia culture. In particular, I explore Fredric Jameson's critique of the 'nostalgia mode', which is still the most oft-quoted theorisation of contemporary nostalgia, and one that I will argue is insufficient.

Chapter 2 explores historically predominant conceptualisations of the relationship between nostalgia, memory and utopia. It focuses on two competing threads in twentieth-century utopian thought. The first consists of the view (arising out of modern ideas about progress) that utopia is, or should be, forward-looking and that, as such, both nostalgia and memory must therefore be inherently anti-utopian. The second conceptualisation is not forward-looking but backward-looking. It resonates with powerful nostalgia either for earlier societies or for an imagined sense of belonging or prelapsarian unity. The utopian, in this view, is aligned with memory; whether on a personal, historical or metaphysical level, utopia is something we have to re-find or re-remember.

In Chapter 3 I present my core theoretical approach, outlining a more promising 'third way' in which memory and nostalgia are neither antithetical to a future-oriented utopian drive, nor sources of a backward-looking utopian longing. I draw on Bloch's critical approach to culture in general, as well as his ideas about non-contemporaneity, to suggest that nostalgic longing can contain seeds of unfinished utopian potential, which Žižek describes as unredeemed ghosts of the past that haunt us – and taunt us – from the side-alleys of history.[59] In other words, I suggest that Bloch might provide us with a way of looking at contemporary nostalgia culture which is more nuanced and indeed more hopeful than prominent critics have allowed. I supplement these core Blochian ideas with comparative reflections on key concepts from Benjamin and Marcuse, intended to enrich the central ideas and highlight particular issues and possibilities.

This third chapter does not attempt to construct a watertight argument to be applied in Part II. This would run counter to the spirit of immanent critique to which Bloch, Benjamin and Marcuse were committed, and would in any case fail to bring out the particular nuances of the kinds of contemporary nostalgia that I consider. However, this chapter does look back to these thinkers and traces in their works ways of understanding the past and the present, nostalgia and hope, that differ from the dominant narratives outlined in Chapters 1 and 2. Thus, I aim to bring to life an intellectual lineage and theoretical perspective which will illuminate the present context, and which may well have much to say about other kinds of cultural production as well – especially regarding how we should read popular and middlebrow cultural works.

In Part II, I will give a reading of several contemporary nostalgia texts that vary in both medium and form. The first is a blog and online webzine by blogger Tavi Gevinson. The second is the 1999 film *The Virgin Suicides* by director Sofia Coppola (and the Jeffrey Eugenides novel from which it is adapted). The third is photography of Detroit's 'abandoned spaces' or urban ruin, with a particular focus on the work of four photographers.

Each of these texts is overtly nostalgic in tone, and is consumed, I argue, as part of the broader nostalgia culture. They are also connected in different ways to one another, thus existing not just as examples of a kind of cultural text, but as interconnected aspects of that culture. Graeme Gilloch, writing of Walter Benjamin, notes:

> His understanding of the way in which the meaning – of a text or historical event – is continually being reconstituted and reconfigured through textual mortification, political appropriation and individual/collective remembrance leads to the fundamental insight that the work of art is legible in specific ways only at particular historical moments – that is to say, as past and present interests intersect, as a critical constellation is formed.[60]

I would argue that this constitution of meaning within a critical constellation is as relevant for popular culture as it is for works of art. My aim is not to attempt to speak directly about contemporary nostalgia culture as a whole, which would assume a static or solid entity about which definitive claims could be made. Moreover, my choice of texts is not intended as a representative or exemplary corpus, in which case it would be a slim corpus indeed. Drawing on the notion of a constellation, I rather offer close and in-depth readings of a particular set of interrelated and constellated texts, in the hope that something will emerge about their truth within the broader culture of their moment, and thus about the broader nostalgia culture of which they are a part. Of course, each of the texts I consider could well have been gathered into different constellations, within which they might have been legible in very different ways. For example, in my reading of *The Virgin Suicides* I look at the film, as well as the novel, and I consider, among other things, their reception in recent years among teenage girls and the meanings ascribed to them within this context. An exploration of Eugenides' work alongside other well-known contemporary American authors such as Jonathan Franzen or Jonathan Safran Foer would have brought out a different set of meanings.

Chapter 4 looks at the media output of teen sensation Tavi Gevinson. Tavi first garnered public attention for her fashion blog, *Style Rookie*. In 2011, when she was nearly fifteen, she then founded *Rookiemag*, a very popular online magazine for teenage girls. The starting point of the chapter is the observation that Tavi's writing and imagery are imbued with nostalgia. The questions it asks are: What is the nature of this temporally blurred, suburban nostalgia? Why has it had such a strong appeal for teenage girls in their cyberspace musings? Is this nostalgia regressive, a sign of cultural inertia, or is there something more hopeful at play?

Chapter 5 gives a reading of Sofia Coppola's 1999 film, *The Virgin Suicides*, an adaptation of the 1993 novel by Jeffrey Eugenides. The chapter considers *The Virgin Suicides* alongside other nostalgia films, and asks whether Jameson's critique of the nostalgia film adequately accounts for the nostalgia of *The Virgin Suicides*. The chapter will explore the 'teenage' as a privileged site of nostalgic longing. It will consider this teenage nostalgia in light of the long-standing association of nostalgia with childhood, and ask: What is the quality of the teenage 'forever' that the nostalgia in *The Virgin Suicides* taps into? What particular hopes are stored therein, and how do they differ from those provoked by childhood nostalgia?

Chapter 6 considers photography of Detroit's 'abandoned spaces', images that are often considered nostalgic in affect. It focuses on the work of four particular photographers, but pays attention also to the wider popularity of 'abandoned spaces' imagery. The chapter explores existing criticisms and defences of these photographic projects of the once great motor city, and asks: Is this photography nothing more than 'ruin porn', subsisting on a nostalgia for Detroit's prosperous Ford-era past? Is it a romanticisation and aestheticisation of poverty by out-of-towners who do not have to live with it? Or can critical and utopian possibility be read in this prolific imagery of industrial abandonment? It will conclude by offering a Blochian reading that overcomes problems faced even by existing Benjamin-inspired readings of this kind of photography.

I am aware that I have not as yet offered the kind of potted summary of my argument that is sometimes given, as a kind of promise to the reader, in introductions of this kind. This is not only a tactic to urge the reader to continue! It is, rather, a reflection of my belief that in matters of affect and culture, the real insights cannot be pithily summarised, but emerge more truthfully in the detail, and nuance, and the bigger picture that detailed reading and careful argument creates. But I can offer some indication of my hopes for the book.

In the spirit of both Bloch and Benjamin, I hope that it will draw attention to the ways in which longings that are out of kilter with a given moment are in some ways the greatest clue to understanding it. I hope that this will have far-reaching consequences for how we view contemporary culture. Should we despair, as Mark Fisher does, that 'the young are no longer capable of producing surprises', dismissing cultural production as less revolutionary or innovative than the cultures of yesterday?[61] Or should we look askance at contemporary cultures? Perhaps hope, and indeed critical resistance – although not always signalling themselves as overtly as they did, for example, in 1960s counterculture – nevertheless are hiding in plain sight, in the nooks and crannies of the very cultures that might seem to have ingested most wholly the maxim 'there is no alternative'.

More broadly still, I hope that this book will demonstrate that a Blochian approach to cultural and literary criticism has a great deal to offer contemporary literary and cultural studies – particularly at a time when the political Left and Right are so starkly divided – and that it can help us to better navigate some of the seeming dichotomies that all too often in recent literary (not to mention political) discourse seem to have calcified: past, future; ideology, utopia; nostalgia, hope.

Notes

1 Fred Davis, 'Nostalgia, Identity and the Current Nostalgia Wave', *The Journal of Popular Culture*, 11.2 (1977), 414–424 (p. 412).
2 Owen Gleiberman, 'Why Nostalgia Movies Leave us Dazed (but Not Confused)', *Entertainment Weekly*, 4 April 2009, www.ew.com/article/2009/04/04/why-nostalgia-m [accessed 12 January 2012]. (Note: although the film was set in 1962 it was before the major cultural and political shifts of the 1960s and so exhibits a typically 1950s-era style).
3 Stephen Marche, '*Dazed and Confused* was the Definitive Movie About the '90s, Not the '70s', *Esquire*, 6 March 2013, www.esquire.com/entertainment/movies/a20266/dazed-and-confused-20th-anniversary-15175511/ [accessed 12 January 2016].
4 Lucy Mangan, '*Stranger Things* Review – A Spooky Shot of 80s Nostalgia Straight To Your Heart', *Guardian*, 15 July 2016, www.theguardian.com/tv-and-radio/2016/jul/15/stranger-things-review-a-shot-of-80s-nostalgia-right-to-your-heart-winona-ryder [accessed 20 August 2016].
5 Mangan, '*Stranger Things* Review'.
6 Mangan, '*Stranger Things* Review'.
7 *Rookiemag* submissions page, http://rookiemag.com/you [accessed 2 June 2013].
8 Paul Grainge, *Monochrome Memories: Nostalgia and Style in Retro America* (Westport, CT: Praeger Publishers, 2002), p. 44.
9 Christine Sprengler, *Screening Nostalgia: Populuxe Props and Technicolour Aesthetics in Contemporary American Film* (New York: Berghan, 2009), p. 30.
10 Mike Chopra-Gant, *"The Waltons": Nostalgia and Myth in Seventies America* (London: I.B.Tauris, 2013), p. 54.
11 Chopra-Gant, p. 55.
12 Leonard Sloane, 'Nostalgia for Extinct Pop Culture Creates Industry', *New York Times*, 22 March 1970, www.nytimes.com/1970/03/22/archives/nostalgia-for-extinct-pop-culture-creates-industry.html?_r=1 [accessed 3 December 2015].
13 Wayne King, 'The Boom in Nostalgia Turns Junk into Junque', *New York Times*, 8 August 1970, www.nytimes.com/1970/08/08/archives/the-boom-in-nostalgia-turns-junk-into-junque-the-boom-in-nostalgia.html [accessed 3 December 2015].
14 Davis, pp. 421–424.
15 Fredric Jameson, '*The Shining*', *Social Text*, 4 (1981), 114–125.
16 Sprengler, p. 70.
17 Sloane, 'Nostalgia for Extinct Pop Culture Creates Industry'.
18 Michael Kammen, *Mystic Chords of Memory: The Transformation of Tradition in American Culture* (New York: Alfred A. Knopf, 1991), p. 656.
19 Georgia Dullea, 'Sick of the 60s, 3 Men of the 80's Try to Give Nostalgia a Bad Name', *New York Times*, 30 April 1989, www.nytimes.com/1989/04/30/style/life-style-sick-of-the-60-s-3-men-of-the-80-s-try-to-give-nostalgia-a-bad-name.html [accessed 4 August 2016].
20 Grainge, p. 42.
21 Hutcheon, p. 192.

22 Sprengler, p. 22.
23 Svetlana Boym, 'Nostalgia and its Discontents', *The Hedgehog Review*, 9.2 (2007), 7–18 (p. 17).
24 Andrew Potter, 'Mad Men and the end of Delicate, but Potent Nostalgia', *National Post*, 15 May 2015, www.nationalpost.com/m/wp/blog.html?b=news.nationalpost.com/arts/television/mad-men-and-the-end-of-delicate-but-potent-nostalgia [accessed 16 May 2015].
25 Touré, 'Nostalgia: Our Favourite Cultural Copout', *Time*, 26 January 2012 http://ideas.time.com/2012/01/26/nostalgia-our-favorite-cultural-copout/?iid=sr-link1 [accessed 3 December 2015].
26 Grainge, p. 44.
27 David Lowenthal, 'Nostalgia Tells It Like It Wasn't', in *The Imagined Past: History and Nostalgia*, ed. by Christopher Shaw and Malcolm Chase (Manchester: Manchester University Press, 1989), pp. 18–33 (p. 29).
28 Katharina Niemeyer, 'Introduction: Media and Nostalgia', in *Media and Nostalgia: Yearning for the Past, Present and Future*, ed. by Katharina Niemeyer (London: Palgrave Macmillan, 2014), pp. 1–23 (p. 1).
29 Hutcheon, p. 192; Kammen, p. 656.
30 Mike Ryan, '"Dazed and Confused" Would Never Work Today', *The Huffington Post*, 10 November 2013, www.huffingtonpost.com/mike-ryan/dazed-and-confused-20-years_b_4085398.html [accessed 12 January 2016].
31 Natalie Olah, 'The Argument Against Nostalgia: Let's Push Things Forward', *I-D*, 29 August 2014, https://i-d.vice.com/en_gb/article/lets-push-things-forward [accessed 16 May 2015].
32 Of course, there was nostalgia long before the 1970s, as readers of Proust or, indeed, *The Great Gatsby* can attest, and the question of how this earlier nostalgia is or isn't different in kind to the nostalgia fuelled by popular cultural reminiscence that has flourished in the last fifty years will be a key concern of this book.
33 Svetlana Boym, *The Future of Nostalgia* (New York: Basic Books, 2001), p. xiv.
34 Susan Sontag, *Against Interpretation and Other Essays* (London; New York; Toronto: Penguin, 2009), p. 311.
35 Fredric Jameson, *Archaeologies of the Future: The Desire Called Utopia and Other Science Fictions* (London: Verso, 2005).
36 Mark Fisher, *Capitalist Realism: Is There No Alternative?* (Ropley: O-Books, 2009), p. 3.
37 Dullea, 'Sick of the 60s, 3 Men of the 80s Try to Give Nostalgia a Bad Name'.
38 Touré, 'Nostalgia: Our Favourite Cultural Copout'.
39 Herbert Marcuse, *Eros and Civilization* (Oxford: Routledge, 1998), p. xviii.
40 Janet Stewart, 'Breaking the Power of the Past Over the Present: Psychology, Utopianism and the Frankfurt School', *Utopian Studies*, 18.1 (2007), 21–42 (p. 27).
41 Stewart, 'Breaking the Power of the Past Over the Present', p. 27.
42 Ruth Levitas, *The Concept of Utopia* (Witney: Peter Lang, 2011), p. 152.
43 Leon Wieseltier, 'Preface', in *Illuminations*, By Walter Benjamin (New York: Random House, 2002), pp. vii–x (p. vii).
44 David Kaufman, 'Thanks for the Memory: Bloch, Benjamin and The Philosophy of History', in *Not Yet: Reconsidering Ernst Bloch*, ed. by Jamie Owen Daniel and Tom Moylan (London: Verso, 1997), pp. 33–53 (p. 48).
45 Wieseltier, 'Preface', *Illuminations*, p. vii.
46 Kaufman, p. 33.
47 Johan Siebers, 'Ernst Bloch's Dialectical Anthropology', in *The Privatization of Hope: Ernst Bloch and The Future of Utopia*, ed. by Peter Thompson and Slavoj Žižek

(Durham, NC: Duke University Press, 2013), pp. 61–82 (p. 61); Vincent Geohegan, *Ernst Bloch* (London: Routledge, 1996), p. 3.

48 Kaufman, p. 35; J.K. Dickinson, 'Ernst Bloch's *The Principle of Hope*: A Review of and Comment on the English Translation', *Babel*, 36.1 (1990), 7–31 (p. 8).

49 Slavoj Žižek, 'Preface', in *The Privatization of Hope: Ernst Bloch and the Future of Utopia*, ed. by Peter Thompson and Slavoj Žižek (Durham, NC: Duke University Press), pp. xv–xx (p. xx).

50 José Esteban Munoz, *Cruising Utopia: The Then and There of Queer Futurity* (New York: New York University Press, 2009); Caroline Edwards, 'Fictions of the Not Yet: Time and the Contemporary British Novel' (unpublished doctoral thesis, University of Nottingham, 2010); Caitríona Ní Dhúill, *Sex in Imagined Spaces: Gender and Utopia from More to Bloch* (Oxford: Legenda, 2010); Johan Siebers, 'The Utopian Function of Film Music', in *Marx at the Movies: Revisiting History, Theory and Practice*, ed. by Ewa Mazierska and Lars Kristensen (London: Palgrave Macmillan, 2014), pp. 46–61.

51 Pedro Erber, 'Contemporaneity and Its Discontents', *Diacritics*, 41.1 (2013), 28–48 (p. 29).

52 Erber, 'Contemporaneity and Its Discontents', p. 29.

53 Kaufman, p. 33.

54 Giorgio Agamben, *What is an Apparatus, and Other Essays*, trans. by David Kishik and Stefan Pedatella (Stanford: Stanford University Press, 2009), p. 40.

55 Agamben, pp. 40–41.

56 Agamben, p. 44.

57 Bloch, *The Principle of Hope*, I, p. 3.

58 Agamben, p. 47.

59 Žižek, Preface, *The Privatization of Hope*, p. xix.

60 Graeme Gilloch, *Walter Benjamin: Critical Constellations* (Cambridge: Polity Press, 2002), p. 234.

61 Fisher, *Capitalist Realism*, p. 3.

Part I

1 Nostalgia

Is it really not what it used to be?

'Nostalgia ain't what it used to be.' This remark was originally a lightly mocking quip at the nostalgic's capacity to make anything, even nostalgia itself, the subject of fond reminiscence. It is often invoked more pointedly by contemporary theorists, who use it to suggest that nostalgia *has* actually changed in recent years or decades, that it now has a different quality.[1] But does it? And if it does, then it what ways, and why? This chapter will provide an overview of influential theories of and ideas about nostalgia via a history of the concept and of literature on the topic. How did nostalgia, originally a term for a specific medical condition, come to be understood as a symptom of the modern age? Why has nostalgia been alternatively described as an anti-modern sentiment and as a peculiarly modern phenomenon? And most recently: What is meant by postmodern nostalgia as a consumer mode? Susan Sontag suggests that 'the two poles of distinctively modern sentiment are nostalgia and utopia'.[2] The story of utopia's rise and fall in modernity is well known, or at least much discussed, if not always agreed on. The story of nostalgia's complex relationship to modernity, and how that has been differently perceived, is less prolifically written about but no less interesting.

A useful starting point is Andrew Higson's account of nostalgia as a changing concept.[3] He outlines three key stages of its development. First, the term described a physical condition provoked by geographical dislocation and a desire to return home; then it came to denote a psychological condition typified by temporal longing for a time gone by; more recently, it has been used to describe a socio-cultural condition that is atemporal, referring not to a longing for the past but to a consumer mode and way of relating to certain styles. Beginning within these broad parameters, I will at each stage draw attention to complexities, inconsistencies and outstanding questions vis-à-vis the status and understanding of nostalgia.

A medical origin: longing for home

The term nostalgia derives from the Greek roots *nostos* and *algia*. *Nostos* means homecoming, and *algia* denotes a painful condition. Interestingly, Svetlana Boym's *The Future of Nostalgia*, a prominent contemporary study of nostalgia, mistakenly translates the Greek term *algia* as meaning longing.[4] The connotation of longing is not in fact present in the original Greek *algia*, which denotes pain, but the fact that Boym's minor mistranslation went unnoticed is perhaps indicative of how central longing has become to the meaning of nostalgia.

The semantic origins of the word *nostalgia* were carefully considered by the Swiss physician Johannes Hofer, who first coined the term in his medical dissertation in 1688. He writes in his introduction:

> Nor in truth, deliberating on a name, did a more suitable one occur to me, defining the thing to be explained, more concisely than the word *Nostalgia*, Greek in origin and indeed composed of two sounds, the one of which is *Nostos*, return to the native land; the other *Algos*, signifies suffering or grief.[5]

The new medical term that Hofer introduced described a kind of acute homesickness. The nostalgic was someone who experienced an intense longing for his native land. The condition was originally associated with Swiss soldiers fighting abroad, although also associated with other displaced peoples, such as students studying abroad and domestic helpers working abroad.[6] As Higson's association of the original term with a 'physical condition' suggests, the person suffering from nostalgia manifested actual physical symptoms – such as fever, nausea and loss of appetite – even though the cause may have been psychological. It is unclear whether Hofer was merely describing and medicalising a long-existing experience, or whether, and to what extent, his naming of the disease coincided with increasing incidences of the experience. Boym suggests that naming it may have 'enhanced the epidemic'.[7] What is clear is that over the course of the eighteenth and early nineteenth centuries nostalgia became well established as a disease. As Roth points out in his study of the medicalisation of nostalgia in nineteenth-century France, 'the disease attracted considerable medical attention, discussion and intervention'.[8]

Modern nostalgia: longing for lost time

By the start of the twentieth century the meaning of the word nostalgia had shifted dramatically. First, it became 'semantically unmoored from its medical basis' and entered into popular vocabularies across Europe as a widely recognised term.[9] The new general concept was no longer associated with physical symptoms, as had characterised the medical definition of nostalgia, but came to describe a purely psychological state. It is not clear exactly when this popular usage of the term gained traction. Heike Jenß states that the meaning of the term shifted in the first half of the nineteenth century and Pickering and Keightley suggest that it happened gradually over several centuries.[10] What is clear is that by the twentieth century, as Hutcheon poignantly observes, this curable medical condition became an incurable condition of the spirit or psyche.[11]

The second significant shift was from spatial to temporal dimensions. Rather than signifying a longing for a missed geographical place, as it had originally, nostalgia came instead to connote a longing towards a missed time. The Oxford English Dictionary points out that the English term *nostalgia* is associated with the German word *Heimweh* or 'homesickness' (Hofer's dissertation was in fact titled *Dissertatio Medica de Nostalgia, oder Heimweh*). The association with a kind of homesickness remained; however, it came to connote 'a sort of homesickness for a lost past' – a temporal dislocation, 'the sense of feeling oneself a stranger in a new period that contrasted negatively with an earlier time in which one felt, or imagined, oneself at home'.[12] Although the retained association with 'home' is significant (this will be further analysed in Chapter 3), there is an obvious but crucial distinction between a spatially distant home and a temporally distant home: the latter, unlike the former, can never actually be returned to. Hutcheon puts it simply: 'Nostalgia becomes the reaction to that sad fact' (the irreversibility of time).[13]

The distinction between spatial and temporal nostalgia is of course not clear-cut. As Higson observes, 'the nostalgic past is almost always a place as well as a time'.[14] Likewise, nostalgia for a place is often associated with a particular time. Hutcheon picks up on an observation made by Kant as early as 1798 that people who did return home were usually disappointed because, in fact, they did not simply desire a lost place, but also a time, a time of youth.[15] However, the new focus on time, rather than place, was nonetheless a qualitative shift which carried with it different resonances. The loss evoked by 'lost time' is a wistful yearning – tinged both by the hopeless impossibility of ever actually going back in time, and the experience in dream or fantasy of having somehow returned. Already widely resonant at the turn of the century, Grainge points out that this new sense of nostalgia was given further voice by 'the triangulation of Freud, Proust and Faulkner' who 'problematized and poeticized the experience of melancholy, memory and nostalgia'.[16] The modern notion of nostalgia, imbued as it is with a poetic and philosophical longing, had fully emerged.

Nostalgia versus modernity

The understanding of nostalgia as anti-modern has a long history and is related to the deep and long-standing association between modernity and progress. The ideal of progress emerged during the Enlightenment, notably in Immanuel Kant's famous 1784 essay, *What is Enlightenment?*[17] It has since come to feature in our understanding of what is meant by modern so prominently that many theorists and historians have suggested that a normative concept of progress is at the heart of the definition of modernity. Immanuel Wallerstein, for example, writes that the term modern 'was situated in a conceptual framework of presumed endlessness of technological progress, and therefore of endless innovation'.[18] More specifically, Pickering and Keightley point out that classical sociology assumed a complete rupture between modern and premodern societies and that the present was positively valued over the past as progress.[19]

Given that nostalgia is enamoured with the past, it is not surprising that it was 'negatively othered as progress's opposite'.[20] If modernity is characterised by future-orientation and progress, then arguably anything that is past-oriented is by default anti-modern. Bruno Latour emphasises this point:

> The adjective 'modern' designates a new regime, an acceleration, a rupture, a revolution in time. When the word 'modern', 'modernization', or 'modernity' appears, we are defining, by contrast, an archaic and stable past. Furthermore, the word is always being thrown into the middle of a fight, in a quarrel where there are winners and losers. Ancients and Moderns.[21]

In addition to the inherent resistance to nostalgia in the idea of modernity, David Gross gives a compelling historical account of the assault on memory in the late modern era (here the term late modern is intended to denote a historical period, from the turn of the twentieth century onwards, rather than a normative ideal).[22] He describes how, from prehistoric societies until at least the seventeenth century, remembering was considered a positive act, 'the rememberer' held up as an ideal.[23] From the seventeenth century on, however, memory underwent a negative re-evaluation. In part, Gross argues, this was simply a function of the increasing speed of change, which made lessons from the past less relevant to the future than they had been in slower, or unchanging societies: 'When one's world has been significantly transformed, the past does not appear to have much to say to the present.'[24] However, in the mid to late nineteenth century, in addition to this general shift, an active and focused attack on memory emerged from within both the sciences and the humanities. Not only had the reliability of memory come into question in psychology, but, in alignment with the increasing value placed on progress, Gross contends that doubts were raised about the value of and motives for remembering:

> The idea arose that the rememberer sometimes remembers out of fear; an inability to 'face up to' reality that represented a kind of running away – a withdrawal into nostalgic reflection that was suitable only for timid or fainthearted souls more interested in searching for solace in the past than in bravely confronting the world around them.[25]

Both 'the rememberer' and 'the nostalgic' were thus conceived as divergent from or even opposed to modern man whose proper focus was on the future and making progress.

In addition to the negative connotations implied of nostalgia by comparison with the value placed on progress, nostalgia can also be thought of as anti-modern insofar as it evolved as a reaction against some of the experiences and alienation that modern life entails. Nostalgia was conceived of as anti-modern not only because it looked backwards and was therefore antagonistic to the self-perception of the 'modern' as progress-oriented, but also because the feeling of nostalgia arose in part as a response to actual changes in society taking part during the historical period known as modernity. Although scholars disagree as to the exact dating of the modern period, there is agreement about some of the changes that took place over the course of several centuries (industrialisation, advances in global capitalism, technological advances, and the shift from 'community' to 'society' described by Tönnies).[26]

Theorists of nostalgia often cite these changes as an explanation for the emergence of nostalgia as a widely experienced and well understood state of mind. Pickering and Keightley, for example, describe nostalgia as 'a reaction to the velocity and vertigo of modern temporality'.[27] Understood as a resistance to the rapid changes of the modern period, nostalgia can be said to embody a rejection of the constant call of progress and transience, and a longing towards slower, steadier and more ontologically secure rhythms. Ralph Harper, for example, explores the existential appeal of nostalgia in the modern age via a reflection on the Grimms' fairy tale, 'Sleeping Beauty'.[28] In a similar vein, Davis argues that nostalgia seeks to secure us against the discontinuities that modernity brings with it, and the accordant existential threat, by appealing to the desire for continuity:

> It is the threat of identity discontinuity (existentially, the panic fear of the 'wolf of insignificance') that nostalgia, by marshalling out psychological resources for continuity, seeks to abort, or, at the very least, deflect.[29]

As Tannock argues, however, while nostalgia may arise out of a desire for continuity in response to change or discontinuity, it also posits discontinuity as part of its own narrative. In the rhetoric of nostalgia one finds three key ideas: a prelapsarian world (the Golden Age, the childhood home, the country); a lapse (a cut, a catastrophe, a separation or sundering, the Fall); and a present, postlapsarian world (a world felt to be lacking, deficient or oppressive).[30] In other words, although nostalgia resists the speed of change in modernity, and on some level desires continuity or stability, it ironically works as well to further heighten that experience of change, of a sharp distinction between 'then' and 'now'.

The resistance to modern temporality and intensity that nostalgia typifies brings to mind the artistic and literary movement that perhaps most overtly speaks in nostalgia's name, Romanticism. Indeed, Jameson's description of Romanticism could equally be applied to nostalgia:

> Romanticism was only new unintentionally: it may indeed be thought of as the way a whole generation attempted to shelter itself, as an organism wards off shock, against that stupendous, total and unprecedented transformation of the world into the henceforth barren and materialistic environment of middle-class capitalism.[31]

Finally, several theorists argue that nostalgia emerged not only in resistance to social and cultural changes brought about by industrialisation, but also to underlying shifts in worldview wrought in modernity. Boym points towards the loss of a stable worldview, *Weltanschauung*, that explained the 'totality of existence', as an important object of nostalgic longing for the fragmented modern.[32] More explicitly, George Steiner talks about a moral and emotional emptiness left in the wake of the decline of formal religious systems, and how this cued a longing for the security of the known which he calls 'nostalgia for the absolute'.[33] Along similar lines, Georg Lukács offers the well-known idea in his early theory that in modernity man is 'transcendentally homeless'. This idea is evident in his elegiac opening to *The Theory of the Novel* (1916):

> Happy are those ages when the starry sky is the map of all possible paths – ages whose paths are illuminated by the light of the stars. Everything in such ages is new and yet familiar, full of adventure and yet their own. The world is wide and yet it is like home, for the fire that burns in the soul is of the same essential nature as the stars.[34]

The 'home' that the modern nostalgic longs for, the 'home' that is lost in time is, according to these theorists, in part a transcendental home; and the nostalgic is explained not only as a figure responding to new and unsettling ways of being and experiencing society and time, but also to transcendental displacement, the severance of cosmological belonging.

James Phillips argues that 'Odysseus longs for home; Proust is in search of lost time'.[35] Although Odysseus's return was not to a home unchanged (he brought new wisdom from his journey which changed both him and the home to which he returned), the original readers of *The Odyssey* could rest assured that the moral order of the universe was permanent, and that their only task was to discover it. Proust's readers, by contrast, have no such guarantee. The 'lost time' that Proust gives voice to is on some level also a time of transcendental belonging, the 'transcendental home' to which one can never return.

These arguments all position nostalgia as in some way anti-modern. Some of these arguments are semantic: nostalgia is necessarily pitted against what is 'modern' if modernity is conceptualised as inherently tied to progress. Others are historical: nostalgia as we know it today (a longing for lost time) emerges strongly in the nineteenth century as a response to the hitherto unprecedented and rapid changes in society, technology, culture and life. Still others are metaphysical: nostalgia is not only a response to changes in society but to a more fundamental loss of a worldview in which man was at home, the stars a known and comforting ceiling rather than an unknown expanse beneath which modern man rattles around.

However, if nostalgia is a reaction to modernity, it is also inherent to modernity. As already noted, Susan Sontag refers to nostalgia as a 'distinctively modern sentiment', alongside utopia.[36] Although nostalgia resists a purely positive evaluation of modernity, and longs for that which is hidden or lost in the modern era, it is only in the modern era that such a longing need arise. As such, nostalgia is both anti-modern and yet a peculiarly modern phenomenon. This is why Boym suggests that nostalgia is 'a symptom of our age'; more than this, even, that it is 'the incurable modern condition'.[37] If progress is the poster-child of modernity, then nostalgia might be described as its shadow, its symptom, its error. It is no less, however, a function of modernity:

> It is not necessarily opposed to modernity and individual responsibility. Rather it is coeval with modernity itself. Nostalgia and progress are like Jekyll and Hyde: alter egos.[38]

Modern technologies of nostalgia

An important line of thinking argues that nostalgia is not just a reaction to the changes wrought by industrialisation and new technologies, but is actually enhanced by those very technologies. There is an irony to this: the nostalgic fears technology, but it is via new technologies and new media that nostalgia is given its most poignant expression.

In particular, the invention, development and mass marketing of photography shaped the experience of memory and nostalgia in far-reaching ways. As a 'snapshot' of the reciprocal relationship between the proliferation of photographic imagery in the twentieth century and the experience of nostalgia, Jenß draws attention to a scene in *Mad Men* in which Don Draper presents a new slogan to his client, Kodak – '*Nostalgia, it's delicate, but potent*':

> In a darkened room with the clicking and sliding round of the Carousel, he displays private photos of his family, his pregnant wife, his cheerful daughter and son – moments of time gone by and yet still there, circulating in images and memories, open to retrospection and revision.[39]

This scene is well chosen on many levels. It viscerally resonates with how photography can feed into and foster nostalgic feelings, and it makes a link to Kodak's 1950s advertising campaign – pointing to the active promotion of nostalgia, as well as the inherent possibilities for nostalgia that the photograph entails, in capturing a slice of time gone by. On another level, the series *Mad Men* is itself highly nostalgic, and the 'clicking' and 'sliding' of what now seems like old-fashioned technology works to induce nostalgia in the contemporary viewer.[40] The origin of photographic nostalgia is thus linked to our meta-experience of it today, showing just how prolific mediated experiences of nostalgia have become.

The ways in which the photograph itself opens up new and different possibilities for memory and encourages nostalgic rumination are beautifully explored by Susan Sontag in her seminal book, *On Photography* (1977). She observes that the 'camera began duplicating the world at the moment when the human landscape started to undergo a vertiginous rate of change'.[41] As discussed, the modern concept of nostalgia is in many ways a resistance to time's irreversibility, a feeling only accelerated by modernity's ever-increasing momentum. In this climate it is not hard to see how poignant the arrival of a technology which purports to capture a moment in time was and is. Sontag's description of this coincidence is worth quoting:

> It is a nostalgic time right now, and photographs actively promote nostalgia. Photography is an elegiac art, a twilight art. Most subjects are, just by virtue of being photographed, touched with pathos. An ugly or grotesque subject may be moving because it has been dignified by the attention of the photographer. A beautiful subject can be the object of rueful feelings, because it has aged or decayed or no longer exists. All photographs are *memento mori*. To take a photograph is to participate in another person's (or thing's) mortality, vulnerability, mutability. Precisely by slicing out this moment and freezing it, all photographs testify to time's relentless melt.[42]

That a photograph can enhance nostalgia is explicable not only because photographs remind us of lost moments, but because the surrounding emotions that a photograph can evoke are also well suited, almost primed, to the receptive longing of the nostalgic. The object in a photograph is simultaneously present (in the image in front of us) and not-present. Likewise, the object of nostalgia is simultaneously present (in our imagination) and not-present. Both therefore trigger a complex in-between state of dreams, desire and impossibility. Sontag's notion that a photograph lays claim to another reality (quoted in full below) is reminiscent of Boym's assertion that 'the alluring object of nostalgia is notoriously elusive', that the nostalgic yearns for St Elsewhere.[43] The response to viewing a photograph, much like nostalgic remembrance, says something about the imaginations and longings of the present viewer. The physicality of a photograph triggers and enables such reverie, as another rueful passage from Sontag's *On Photography* infers:

> A photograph is both a pseudo-presence and a token of absence. Like a wood-fire in a room, photographs – especially those of people, of distant landscapes and faraway cities, of the vanished past – are incitements to reverie. The sense of the unattainable that can be evoked by photographs feeds directly into the erotic feelings of those for whom desirability is enhanced by distance. The lover's photograph hidden in a married woman's wallet, the poster photograph of a rock star tacked up over an adolescent's bed, the campaign-button image of a politician's face pinned on a voter's coat, the snapshots of a cab driver's children clipped to the visor – all such talismanic uses of photographs express a feeling both sentimental and implicitly magical: they are attempts to contact or lay claim to another reality.[44]

The nostalgic is someone 'for whom desirability is enhanced by distance', the distance of lost time, so it is not surprising that the photograph should feed into the desires (erotic feelings) of the nostalgic.

Andreas Huyssen argues that the forms memory takes are culturally contingent: how a culture experiences temporality, how it relates to both past and future and how it remembers are not everywhere always the same, but are strongly influenced by both the cultural practices and dominant technologies of the time.[45] The photograph is a particularly powerful technology for enhancing the experience of nostalgia, and it is not incidental that the emergence of the romantic kind of modern nostalgia coincides with the development and spread of photography over the nineteenth century.

A subsequent and also significant technological development occurred with the arrival of film. Landsberg argues that the twin developments of mass movement of peoples that industrialisation brought about, and the emergence of mass culture made possible in particular by the cinema, enabled the experience of memories that were not our own personal memories, or those of our own cultural or ethnic group. She calls this new form of public cultural memory 'prosthetic memory'.[46] Her argument is that the emergence of the cinema allowed people to be viscerally transported into experiences of lives that are not their own, or even those passed down within their own culture or community. The upshot of these developments, she claims, was the development in the twentieth century of memories which do not belong to any one group, but which anyone can partake in. For example, she says that 'memories of the Holocaust don't belong only to Jews, nor do memories of slavery belong solely to African Americans'.[47] She recognises that several theorists are anxious about what these burgeoning technologies have meant for authentic memory.[48] This fear – that the commodification of mass culture had a negative impact on memory – was most famously articulated by Adorno and Horkheimer in *The Culture Industry*.[49] But Landsberg argues counter to such fears that the development of 'prosthetic memory' had positive effects as well, primarily the ability to relate on a deeper and more empathetic level to past experiences that are not our own.

The general supposition that I have put forward is that new technologies, particularly photography, had a radical impact on the experience and spread of nostalgia in the twentieth century. Before moving on, I want first to raise a significant counterpoint to this view, which will be more fully developed in Chapter 3. A photograph is usually something that we have to hand and can choose to look at, or dwell on, at will. Indeed, people often do return to old photograph albums when they are in a nostalgic mood. There are those, however, who believe that it is not our voluntary recollections that evoke the most powerful nostalgia, but our 'mémoire involuntaire' – so termed by the most renowned chronicler of this kind of sudden memory, Marcel Proust. Gross points out that for Proust, it was the jarring differences and discontinuities highlighted by sudden and involuntary memory that made them so valuable.[50] Walter Benjamin is another thinker for whom what gives memory its power is the dissonance with present-day needs or desires. It is, as will be explored subsequently, this gap which enables Benjamin to posit a critical function for memory. However, excepting when an old photograph is stumbled upon in error, photographs do not trigger accidental and forgotten memories; they are, rather, sought out to affirm and ruminate on our present memory of a moment. That memory is in turn influenced by the photograph. It is a very familiar experience for people today that a memory of, for example, a past holiday, comes over time to be identical with the photographs taken of it – the un-captured moments and associated feelings that one might otherwise have stumbled upon having been eclipsed by the more readily available photographic evidence. While photography can be said to have enabled or encouraged nostalgia, it is worth pausing to consider more specifically what kinds of memory or nostalgia it has promoted, and what kinds have been lost in its wake.

Postmodern nostalgia: what longing?

By way of brief summary and reminder, Higson posited first a medical (spatial) nostalgia, then a modern (temporal) nostalgia and finally a postmodern (atemporal) nostalgia. The shift from nostalgia as a medical term denoting intense longing for a missed geographical home to the more widespread modern concept of nostalgia as a deep longing for lost time has been discussed above; and the complexities of the modern concept of nostalgia as both a resistance to, but also a product of, modernity have been drawn out. Grainge observes that until the 1970s nostalgia was primarily a philosophical or literary concern (as is appropriate vis-à-vis the modern sense of nostalgia as a poetic, philosophical, longing), and that only in the 1970s did the sociocultural dimensions of nostalgia start to receive more attention.[51] The third stage, or type, of nostalgia, postmodern nostalgia, is most famously articulated by Fredric Jameson in the 1980s and aligns with this shift from the psychological or philosophical to the sociocultural.

Jameson started writing about nostalgia culture early in the 1980s, first in a 1981 essay about Stanley Kubrick's film *The Shining* (1980), but most famously in his 1984 article 'Postmodernism, or The Cultural Logic of Late Capitalism', which he later developed further and published in 1991 as a book.[52] Like others in the 1970s and early 1980s (see Introduction), he observes a proliferation of nostalgia in American culture which he calls 'contemporary "nostalgia" culture, what the French call *a la mode retro*'.[53] He is highly critical of this nostalgia culture in general, and his most developed critique is directed at the 'nostalgia film' and a 'new aesthetic mode', which he called the 'nostalgia mode', and which he saw at play in a number of films. However, this critique of the nostalgia mode is easily applicable, and indeed has been applied both by Jameson and others since, to a much wider variety of contemporary texts and genres than just film (novels, fashion, architecture, music, online media). Jameson's critique of the nostalgia mode was part of a broader critique of postmodern culture as a whole.

Jameson's famous critique of postmodern culture posited postmodernism as the new cultural dominant: not one cultural style among others, but the cultural logic that typified most then-contemporary cultural products (with instances of divergence identifiable by their departure from this hegemonic norm).[54] Crucially, he contrasts postmodern culture with the previous significant cultural moment, modernism (understood here to refer to the literary and cultural movement of the early twentieth century, not the related sociological concept of 'modernity'). It is unnecessary here to give an overview of Jameson's theory of postmodernism as a whole, but the contrast he makes between postmodernism and modernism is important to understanding his view of the nostalgia mode (postmodern nostalgia) as expressly distinct in quality and kind from the nostalgia expressed in modernism, which Jameson understood as similar in quality to the poetic, philosophical 'modern' nostalgia that has already been discussed. As he says when introducing his characterisation of the nostalgia mode:

> Nostalgia does not strike one as an altogether satisfactory word for such fascination (particularly when one thinks of the pain of a properly modernist nostalgia with a past beyond all but aesthetic retrieval).[55]

Grainge clarifies the distinction already present in Jameson and makes it more explicit, by contrasting Jameson's 'nostalgia mode' with the preceding 'nostalgia mood'. As already indicated, modern nostalgia, or the nostalgia mood, is arguably a response to the upheavals and discontinuities of modernity, which searches for stability or authenticity in the past, sometimes in the form of personal memory, particularly of childhood, and sometimes in the form of cultural memory of a 'golden age'. But what is postmodern nostalgia, or the nostalgia mode, and how does it differ from this more familiar conception of nostalgia as a psychological longing for lost time?

Amnesiac nostalgia

The first way of distinguishing the nostalgia mode from its predecessor, the nostalgia mood, is by exploring their different relationships to history and to time. Jameson suggests that modernism reflected an intense fascination with the question of time and temporality, and also therefore with history and with memory, and that this fascination has been eclipsed in postmodernity. Likewise, the modern experience of nostalgia – a Proustian 'search for lost time', imbued as it is with a deep, melancholic yearning and resistance to modernity's irreversible march forward – is, according to Jameson, superseded in late capitalism by the nostalgia mode, which is characterised by a comparatively superficial engagement with the past.

One of the core films in which Jameson identifies and analyses the nostalgia mode is *American Graffiti* (1973). He argues that this film trades on nostalgia, using aesthetic techniques and filters to convey a general sense of pastness and pseudo-historical depth, without containing any actual historical content. Instead of attempting to portray the lived realities of a given historical moment, or to understand how it emerged out of preceding moments or led to others, this kind of nostalgic cultural production presents only a superficial, stylised version of the past. Cultural products like *American Graffiti* approach the past 'through stylistic connotation, conveying "pastness" by the glossy qualities of the image, and "1930s-ness" or "1950s-ness" by the attributes of fashion'.[56] History is thus turned into pastiche, an affectless and random collage of dead styles. The decade of the 1950s, for example, becomes about red Chevrolets, Marilyn Monroe's white dress and rock 'n' roll, rather than about ongoing racial and gender oppression, urban degradation in the wake of post-war suburbanisation or the Korean War. Jameson calls this 'libidinal historicism'. Real history is replaced by the history of aesthetic styles, so that 'we are condemned to seek History by way of our own pop images and simulacra of that history, which itself remains forever out of reach'.[57] A number of other critics agree with Jameson's view of contemporary nostalgia culture. For example Dick Hebdige writes:

> The past is played and replayed as an amusing range of styles, genres, signifying practices to be combined and recombined at will. The then (and the there) are subsumed in the now. The only history that exists here is the history of the signifier and that is no history at all.[58]

The proliferation of the nostalgia mode is seen by Jameson as a function of what he perceives as a wider amnesiac culture. The idea that capitalist culture suffers from a kind of collective amnesia was first articulated by Theodor Adorno in the early post-war period. Adorno's now famous statement in a letter to Benjamin that 'all reification is a forgetting' means, specifically, that to reify something is to view it independently of the process of historical mediation which constitutes it.[59] This entailed the assumption that someone or something is, simply, essentially, what it is, regardless of social or historical context. Reification, which Adorno saw as inherent to the logic of late capitalism, was therefore a kind of historical forgetting. Reification also robs individuals of their historical consciousness, their sense that the world as it is today is contingent rather than necessary, and that it could yet be otherwise.

Fredric Jameson builds on this intellectual background in his own work, which posits that one of the defining features of postmodernism is a loss of connection to history, and therefore of the sense that things could be other than they are. He argues that information overflow, saturation of media imagery and ever-accelerating consumption make a real sense of historical continuity nearly impossible; instead we consume thirstily the impoverished version of pseudo-history that the nostalgia mode presents. Thus the nostalgia mode that proliferates in American popular culture from the 1970s onwards is, for Jameson, an expression of a deeply amnesiac culture which can only remember de-contextualised, de-historicised, superficial imagery and fashion. Ironically, since nostalgia is normally associated with memory, the nostalgia mode is for Jameson premised on forgetting.

Consumption and longing

Another important distinction that Grainge makes is that the nostalgia mood is defined by the experience of loss, by 'crippling reservoirs of pathological longing', whereas the nostalgia mode is notable for its relative lack of loss and longing.[60] The intense longing of the nostalgia mood is alternatively directed at the lost experience of childhood or bygone ways of life. It arguably even expresses an unconscious desire for transcendental belonging, Steiner's 'nostalgia for the absolute'. While the historical accuracy of the perception of the lost object can be questioned (were the bygone ways of life really so desirable?) no one doubts the real feeling of loss that this kind of nostalgia expresses. By contrast, Grainge argues, the nostalgia mode, a taste for styles which connote 'pastness' and which can be consumed or enjoyed in the present, is not necessarily driven by longing or loss:

> Reducing sentiment and style to a fixed and causal relation can underestimate the way that, as a cultural style, nostalgia has become divorced in contemporary culture from a necessary concept of loss.[61]

He argues that many people make the mistake of assuming that the nostalgia style, or mode, necessarily reflects a deeper longing, but that they are wrong to do so. Davis's explanation of the 1970s nostalgia boom as a response to the upheavals of the 1960s and a desire for continuity of identity is one example that, according to Grainge, assumes deeper reasons for nostalgia culture than may be necessary or indeed true. Grainge argues that a consumer appetite for pastness is often simply a reflection of technological and economic developments. The nostalgia boom of the 1970s and subsequent decades was less a reflection of a culture reeling from discontinuity or loss than a reflection of a culture that was newly able 'to transmit, store, retrieve, reconfigure and evoke the past in specific ways'.[62]

Higson, too, agrees that a crucial distinction between modern nostalgia and postmodern nostalgia is the absence of a sense of loss, or yearning, in the latter. If the nostalgia mood (modern nostalgia) is a wistful, bittersweet, hopeless longing for something lost, Higson points out that the current culture, or business, of nostalgia seems 'surprisingly sweet and not at all bitter'.[63] Reynolds also says about the currently fashionable retro sensibility that it 'tends neither to idealize nor sentimentalise the past, but seeks to be amused and charmed by it'.[64]

Nostalgia for sale

Underpinning all the various critiques of the nostalgia mode is the idea that it is on some level commercial nostalgia: nostalgia for sale. This explains the assumed lack of longing. If nostalgia is effectively for sale in contemporary postmodern culture, then the past is not irrecoverable: one needs only to consume the nostalgia product (buy the piece of retro furniture, watch the nostalgia film, eat the old-fashioned candy) in order to enjoy and in a sense relive the past. Jameson is less explicit than others about the commercialisation of nostalgia, focusing more on nostalgia as an aesthetic mode, but his underlying supposition is clear, as for example when he explains the emergence of nostalgia films in the broader context of Hollywood market mechanisms: 'blockbuster film productions now become tightly linked to bestsellers and developments in other branches of the culture industry'.[65]

Arjun Appadurai is even more explicit about the link between commercial interests and the spread of nostalgia culture. He talks about the emergence of what he calls 'armchair' or 'ersatz' nostalgia: 'nostalgia without lived experience or collective historical memory'.[66] This concept is similar to Landsberg's 'prosthetic memory', although Appadurai is much more damning of the phenomenon than Landsberg is. Rather than seeing this kind of memory of what has not been personally experienced as an accidental product or byproduct of new technologies and what they make possible, Appadurai thinks that such ersatz nostalgia is more often an active and intended result of advanced marketing and mass merchandising. Adorno and Horkheimer argued that 'real life is becoming indistinguishable from the movies'.[67] Likewise, Appadurai argues that real memory is becoming indistinguishable from marketed memories:

> Rather than expecting the consumer to supply memories while the merchandiser supplies the lubricant of nostalgia, now the viewer need only bring the faculty of nostalgia to an image that will supply the memory of a loss he or she has never suffered.[68]

West also contends that it is not the technologies themselves, but the corporate interests behind them, that have more than anything shaped our experience of memory and nostalgia.[69] She argues, against Sontag's supposition that there is something desirous and sentimental in the nature of the photograph itself, that the company Kodak actively taught Americans to experience their memories in this way, as objects of nostalgia. For example, West points out that in the 1840s and 1850s post-mortem photographs were very popular. She takes this as evidence that sorrow was allowed into the space of the domestic photograph in a way that, since the onset of the nostalgic snapshot culture promoted by Kodak, it no longer is. The scene from *Mad Men* cited earlier lends credence to this perspective, showing how Kodak actively promoted nostalgia in its advertising campaign.

The nostalgia mode can be used to further not only economic interests, but also political agendas. The most obvious, or at least the earliest, example of this kind of nostalgia mode fuelling politics is Ronald Reagan. Grainge makes the case for this quite strongly:

> No treatment of nostalgia as a cultural style in contemporary America could fail to mention the significance of Reagan. He developed a particular

> form of neo-kitsch, feeding on the past and the present in ways that transmogrified popular culture into a compendium of national motifs. [...] If postmodernity is defined by the cannibalisation of past styles, Reagan became the cannibal-in-chief.[70]

Grainge suggests that Reagan's previous career as a radio, film and television star gave him an 'iconic capital' as part of the 'popular culture past, a symbolic cachet that he was able to build on in creating and playing on a mythic conception of a small-town American past'.[71] The recent TV series *Fargo*, set in 1979, has an episode in which Reagan visits Luverne, Minnesota, on his Presidential campaign, which explores Reagan's intentional use of nostalgia. The actor playing Reagan, Bruce Campbell, commented in an interview about his role: 'I don't know how he did it, but boy he sure made life simple. "Things can be good again. Things can be great again." And the '80s became the '50s again in a weird way.'[72] The parallels with Donald Trump's 2016 slogan 'Make America Great Again' of course indicate that the political deployment of nostalgia is by no means a tool of the Reagan era exclusively.

Technology and the nostalgia mode

Although I drew on Susan Sontag to suggest that the technology and spread of photography in the twentieth century contributed to the development of modern nostalgic longing (the nostalgia mood), the photograph is also seen as important by Jameson in his depiction of postmodern nostalgia. The nostalgia mode reconstructs the past as nothing more than a vast collection of images, which is clearly aided by the proliferation of photographic imagery. Previously I described how a photograph can evoke a powerful and sentimental longing for lost moments, but it is also clear, seen from a different angle, how photographs can reduce the past to nothing other than a series of dislocated snapshots. What evokes memory can also, at its extreme, invoke amnesia. Sontag is also aware of the potential for this, as indicated in the passage below, which clearly resonates with Jameson's description of the 'nostalgia mode':

> Through photography, the world becomes a series of unrelated, freestanding particles; and history, past and present, a set of anecdotes and *faits divers*. The camera makes reality atomic, manageable and opaque. It is a view of the world which denies interconnectedness, continuity, but which confers on each moment a character of mystery.[73]

Siegfried Kracauer, in his well-known essay *Photography*, regrets that 'in a photograph, a person's history is buried as if under a layer of snow'; and West, too, gives further support to this point:

> Modern culture has come to regard the isolated moments represented by photographs as producing collective truth when in fact they represent at

best only relative truths and their sheer proliferation negates meaning rather than helps construct it.[74]

The photograph, then, can either be understood as aiding memory and nostalgic longing, or alternatively as aiding the bite-sized consumption of pieces of the past that trivialise and obscure memory.

We now live in the digital age – the internet, digital photography, digital music – which raises its own set of questions vis-à-vis contemporary production of nostalgia. It is not incidental that Kodak went bankrupt in 2012. The company has since found ways to emerge from bankruptcy, very differently constituted, but the bankruptcy is nonetheless symbolically important. Has the relationship between the photographic image and nostalgia changed in a post-kodakian time?

The collected essays in Niemeyer's *Media and Nostalgia: Yearning for the Past, Present and Future* take a particular interest in new media, and in making sense of the nostalgia boom in relation to digital culture. Several ideas surface in these essays. The first is that the incessant updating and the speed of development of media today means that 'media are very often nostalgic for themselves, their own past, their structures and contents'.[75] Coon makes a similar point, arguing that due to faster cultural turnover 'instead of reminiscing about styles, behaviours and customs a hundred years in the past, we nostalgically remember cultural icons, catch phrases and electronic gadgets from less than five years ago'.[76] Two things are notable: the recent timeframe of the nostalgic object, and the fact that the object of nostalgia is not life itself but older media. This seems to lend credence to a Baudrillardian picture of the real slipping further and further from reach, nostalgia feeding on an endless loop of simulacra, which will be explored in Chapter 4. A good example of this is the TV drama, *Stranger Things*, mentioned in the Introduction. The programme exhibits not just nostalgia for an earlier era, nor even just for earlier filmmakers, but also for the technologies they had at their disposal, and the experience of time that these earlier technologies embodied: 'walkie-talkies as the bleeding edge of technology, [...] lost spirits communicating through mains electricity and crackling phonelines'.[77]

Another point of interest in the Niemeyer collection is the use of nostalgia aesthetics in very recent technologies. For example, Gil Bartholeyns explores the idea of retro digital photography apps that make photographs of the present appear old, like pre-digital photography.[78] This, he argues, creates a kind of 'instant nostalgia', using an aesthetics of the past to nostalgise the present, a way of engaging with the present moment as though it were already a memory. Relatedly, Guiseppina Sapio asks why we shoot contemporary family videos in old-fashioned ways.[79] Another related example, not explored in the collection, is the renewed popularity of the polaroid camera at a time when digital photography is more advanced than ever. There is an irony in all of this: we can now take images, even with iPhone cameras, which are of a higher quality than at any time in history, and yet that advanced technology is so often used to make pictures look old.

Another important and relevant feature of the internet is that the sheer proliferation of imagery and the extent of access to materials of the past lend themselves to exactly the kind of de-contextualised engagement with history that critics bemoan in the nostalgia mode, as does the amorphous structure of the internet (a reader of a book goes from page to page and a viewer of a film watches in the order the director intended, but a browser of the internet browses in any order they choose: jumping around from page-to-page, image-to-image and text-to-text at will). In other words, the nostalgia mode – already rampant in the 1970s – finds its perfect platform on the internet, particularly on blogs and tumblrs and other sharing platforms. As Hila Shachar writes, 'I've been disturbed, time and again, when I've encountered images of Holocaust memorials, people during war and images from the past being consumed as a pleasing visual aesthetic or consumer "product" on many blogs.'[80]

An example of this arrived in my own Instagram feed recently, from the account of the online Russian contemporary arts and culture magazine, *The Calvert Journal*. The magazine's Instagram account, like the magazine itself, presents beautifully shot images which could be described as having a neoromantic aesthetic: vast suburban Soviet blocks seem dreamy rather than depressing when covered in a thin film of snow and shot in pinky-blue tones. Alongside images of this kind shot by contemporary photographers appeared, in the Instagram flow, a black-and-white image of young Stalin. Stalin looks handsome in the photograph, alluring: a checked neckerchief around his neck lends him an air of activist student rebellion, his hair is thick and luscious and the several-day-old stubble on his face would not be out of place in a Calvin Klein advert. The comment reads:

> Photograph of a young #Stalin on display at the 'must see' museum that pays tribute to the pre-eminent tyrant. More memorabilia in the journal #youngdictator #throwback #gori #stalinmuseum.

One excited reader comments in response, simply, 'OooooiiiuuuuuAAAuauuuuuoioioiiiiiiiUUUUhhhhhh!' Of course, readers of a Russian literary and cultural journal can all be expected to know the history of Stalin's leadership, and so can be assumed not to approach the image in a historical vacuum. Nonetheless, the romanticised consumption of the standalone image in the flow of other appealing, unrelated, images is exactly the kind of superficial, stylised engagement with the past that Jameson so decried, and which the internet has only heightened and intensified.

Criticisms of theories of the nostalgia mode

Hutcheon points out that Jameson's own rhetoric and position is strangely nostalgic.[81] In criticising the contemporary nostalgia mode as more superficial than modernist nostalgia he looks to the past as the place of authentic, genuine, real historical engagement. It should first be said that Jameson's nostalgia does not warrant automatic criticism. If the nostalgia mode is a genuinely new, or recent, phenomenon which reflects a more superficial engagement with the past than the nostalgia mood did, then perhaps these critics are right to feel nostalgic for those earlier forms of nostalgia. Mark Fisher makes this point convincingly:

> It seems strange to have to *argue* that comparing the present unfavourably with the past is not automatically nostalgic in any culpable way, but such is the power of the dehistoricizing pressures of populism and PR that the claim has to be explicitly made.[82]

However, one can ask whether Jameson is right to compare the nostalgia mode unfavourably with the nostalgia mood. Indeed, is the nostalgia mode as different from the nostalgia mood as his argument suggests?

This is a central question, and one that will be explored further via the readings of contemporary nostalgia texts in Part II of this book, but some preliminary points can be raised here. Even if the quality of the nostalgia mode is different in kind to the nostalgia mood, the assumption that the nostalgia mood was ever unmediated is evidently wrong. There is a tendency to assume a stark division between new media and what came before, with the newest media often receiving the most attention. While it is right to consider the new kinds of engagement opened up by developments in new media, to posit that there was ever an entirely unmediated form of nostalgia is to ignore the ways in which developing technologies such as the photograph played into nostalgia over the preceding centuries, long before any notion of postmodernism or the nostalgia mode appeared on the intellectual scene.

Something else assumed in theories of the nostalgia mode is the absence of deep longing in contemporary consumer nostalgia (as discussed, Grainge makes this explicit). For Jameson, the nostalgia mode is characterised by the triumph of depthlessness, contains little real longing, and is thus entirely distinct in character from the nostalgia mood, driven as it was by deep longing and a sense of loss at time's irreversibility. But is the nostalgia mode really as divorced from the deeper longings and tussles with modernity as these theorists claim? Are the nostalgia mood and nostalgia mode really so distinct? In my analysis of *Style Rookie* and *Rookiemag* in Part II, I hope to show that this these texts – an online teenage girls' blog and webzine – which at a first glance appear to fit neatly into Jameson's description of the nostalgia mode as the 'random cannibalisation of all the styles of the past, the play of random stylistic allusion', in fact express and reflect deep longings akin to the nostalgia mood, longings which contain within them seeds of utopian potential.[83] Although the act of consuming nostalgia products or imagery may be superficial, the question remains as to why we are drawn to such products and images in the first place. What are we looking for there? Or, put differently, why are nostalgia marketing campaigns so successful?

Current debate

Just as nostalgia culture has not expired, so too academic interest in the phenomenon continues to expand. Although each article or book attempts to contribute to the discussion in specific ways, there are a few broad lines of argument that can be identified in recent scholarship.

First, there are many ongoing efforts to expose nostalgia as dangerous, particularly by drawing attention to its use and manipulation by the political right wing. In the American context, this line of argument tends to focus particularly on the 1970s and 1980s, culminating in the Reagan-era nostalgia fest. Coontz's 1992 book, *The Way We Never Were: American Families and the Nostalgia Trap* explores and explodes various nostalgic myths about the bygone American Family that have been propagated in order to serve a conservative agenda.[84] Chopra-Gant's 2013 book, *The Waltons: Nostalgia and Myth in Seventies America,* takes a similarly broad approach, but focuses particularly on 1970s nostalgia television, and the ideological desire for paternalistic values that seemingly harmless feel-good television embodied. Marcus's 2004 book *Happy Days and Wonder Years: The Fifties and the Sixties in Contemporary Cultural Politics* looks at how these postwar decades have been repeatedly invoked by politicians since 1970.[85] Efforts to expose nostalgia as an ideological tool of the Right are not limited to the American cultural landscape. For example, Owen Hatherley's 2016 book, *The Ministry of Nostalgia: Consuming Austerity (Keep Calm and Carry On)* attempts to expose the ubiquitous 'Keep Calm and Carry On' posters, and accordant mentality, as an effort by the Right in the UK to invoke a utopian past that never existed and thus gain support for their austerity agenda.[86]

Then, there are efforts to rescue nostalgia from its ongoing conception as reactionary and regressive, to unearth critical potential in that which has typically been dismissed as ideological. These can be divided into those which engage seriously with Jameson's critique of the nostalgia mode and those which do not. Approaches of the latter kind tend to proceed as though the nostalgia that needs defending is the same old nostalgia that has long shadowed utopia, that long-standing resistance to the ideal of 'progress' embodied in most definitions and experiences of modernity. They do not seriously engage with the criticisms put forward by Jameson and taken further by others: that the nostalgia mode is something altogether new, a widespread consumption of pastness that is characterised less by deep longing than by superficial enjoyment.

The most well known of these efforts is Svetlana Boym's 2001 book *The Future of Nostalgia,* which is referred to in nearly all recent work on nostalgia. Her theory centres on a distinction between 'restorative' and 'reflective' nostalgia, which I will elaborate on and discuss in Chapter 3. Her work has useful parallels with this book in that she attempts a defence of certain kinds of nostalgia, and at one point even mentions that nostalgia can be seen as having a utopian dimension ('reflective nostalgia has a utopian dimension that consists in the exploration of other potentialities and unfulfilled promises of modern happiness'[87]). This idea, which she mentions only briefly, will be developed in this book via a thorough analysis of the question of what unfulfilled utopian potential means in relation to nostalgia. Boym's work is also pertinent to my own in that she, too, makes use of the ideas of Frankfurt School theorists, particularly Walter Benjamin, to enrich her own analysis. However, she fails to engage in any meaningful way with Jameson's critique of the nostalgia mode, and as such her analysis leaves some major questions unanswered, as will be explored in Chapter 3.

The third major line of argument in the current literature does engage with Jameson's identification of the nostalgia mode, and agrees with some of his characterisation of it, but attempts a defence of it on different grounds. An example of this is Grainge's examination of the use of the black-and-white image in American visual culture during the 1990s. Grainge rejects the idea that the nostalgia mode is inherently amnesiac and explores the ways in which it performs a memory function through the stylistic and affective play of 'pastness'.[88] Another example is Higson's focus on consumer reception. He argues that Jameson relies too heavily on the form of a film, or other cultural product, as itself dictating what it is and what work it does, whereas he wants to allow space for consumer identity, taste and choice, not just textual analysis.[89] This is an important idea, which I will make use of. Other examples of approaches in this category include Dika's argument against Jameson's dismissal of nostalgia films, Jenß's consideration of the performative, critical engagement of the nostalgia style practice of blogger Tavi Gevinson, and Flinn's exploration of the connection between nostalgia, utopia and femininity in Hollywood film music from the 1930s to 1950s.[90]

My own argument, as will become apparent, is aligned with these efforts to save nostalgia from being too easily dismissed as hopeless ideology. Its focus is in particular on challenging the oft-assumed opposition between utopia and nostalgia. The questions at its core are: Can one read hope in contemporary American nostalgia culture? Is nostalgia culture in any way utopian? Are there sparks of critical, utopian potential in contemporary nostalgia texts? In order to approach these questions, I begin, now, by looking at utopian theory. How has nostalgia been understood by theorists of utopia? Can a better understanding of utopia help to shed light on the utopian promise (or lack thereof) of nostalgia?

Notes

1 Andrew Higson, 'Nostalgia is Not What It Used To Be: Heritage Films, Nostalgia Websites and Contemporary Consumers', *Consumption Markets and Culture*, 17.2 (2014), 120–142; Martyn Bedford, 'Nostalgia Ain't What It Used To Be', *Literary Review*, 299 (2003); Grainge, p. 19.

2 Sontag, *Against Interpretation*, p. 311.

3 Higson, p. 123.

4 Boym, *The Future of Nostalgia*, p. xiii.

5 Johannes Hofer, 'Dissertatio Medica de Nostalgia' (Basel, 1688), trans. by Carolyn Kiser Anspach, in *Bulletin of the History of Medicine*, 2 (1934), 376–391.
6 Boym, *The Future of Nostalgia*, p. 3.
7 Boym, *The Future of Nostalgia*, p. 3.
8 Michael Roth, 'Dying of the Past: Medical Studies of Nostalgia in Nineteenth-Century France', *History and Memory*, 3.1 (1991), 5–29 (p. 7).
9 Michael Pickering and Emily Keightley, 'The Modalities of Nostalgia', *Current Sociology*, 54.6 (2006), 919–941 (p. 922); (Davis, p. 415, makes an almost identical point that nostalgia was unmoored from its pathological base).
10 Heike Jenß, 'Cross-Temporal Explorations: Notes on Fashion and Nostalgia', *Critical Studies in Fashion and Beauty*, 4.1–2 (2013), 107–124 (p. 113); Pickering and Keightley, p. 922.
11 Hutcheon, p. 194.
12 Pickering and Keightley, p. 922.
13 Hutcheon, p. 194.
14 Higson, p. 123.
15 Hutcheon, p. 194.
16 Grainge, p. 20.
17 Immanuel Kant, 'Answer the Question: What is Enlightenment?', in *Practical Philosophy*, trans. and ed. by Mary J. Gregor (Cambridge: Cambridge University Press, 1996), pp. 11–22.
18 Immanuel Wallerstein, *After Liberalism* (New York: New Press, 1995), p. 173.
19 Pickering and Keightley, p. 919.
20 Pickering and Keightley, p. 920.
21 Bruno Latour, *We Have Never Been Modern*, trans. by Catherine Porter (Cambridge, MA: Harvard University Press, 1993), p. 10.
22 David Gross, *Lost Time: On Remembering and Forgetting in Late Modern Culture* (Boston: University of Massachusetts Press, 2000).
23 Gross, p. 25.
24 Gross, p. 31.
25 Gross, p. 33.
26 Ferdinand Tönnies, *Community and Society (Gemeinschaft und Gesellschaft)*, trans. by Charles P. Loomis (New York: Harper & Row, 1963).
27 Pickering and Keightley, p. 923.
28 Ralph Harper, *Nostalgia: An Existential Exploration of Longing and Fulfilment in the Modern Age* (Cleveland, OH: The Press of Western Reserve University, 1966).
29 Davis, p. 420.
30 Stuart Tannock, 'Nostalgia Critique', *Cultural Studies*, 9.3 (1995), 453–464 (pp. 456–457).
31 Fredric Jameson, *Marxism and Form: Twentieth Century Dialectical Theories of Literature* (Princeton: Princeton University Press, 1974), p. 94.
32 Boym, *The Future of Nostalgia*, p. 24.
33 George Steiner, *Nostalgia for the Absolute* (Concord, ON: House of Anansi Press, 1997).
34 Lukács, Georg, *The Theory of the Novel*, trans. by Anna Bostock (Cambridge, MA: MIT Press, 1968), p. 29.
35 James Phillips, 'Distance, Absence and Nostalgia', in *Descriptions*, ed. by Don Ihde and Hugh J. Silverman (Albany: SUNY Press, 1985), p. 65.
36 Sontag, *Against Interpretation*, p. 311.
37 Boym, *The Future of Nostalgia*, p. xvi, p. xiv.
38 Boym, *The Future of Nostalgia*, p. xvi.

39 Jenß, p. 108.
40 This episode of *Mad Men*, 'The Wheel' aired on 18 October 2007. Kodak filed for bankruptcy in 2012, lending the episode, if watched now, additional nostalgic poignancy.
41 Susan Sontag, *On Photography*, (London: Penguin, 2008), p. 16.
42 Sontag, *On Photography*, p. 15.
43 Boym, *The Future of Nostalgia*, p. xiv.
44 Sontag, *On Photography*, p. 16.
45 Andreas Huyssen, *Twilight Memories: Marking Time in a Culture of Amnesia* (New York: Routledge, 1995), p. 2.
46 Alison Landsberg, *Prosthetic Memory: The Transformation of American Remembrance in the Age of Mass Culture* (New York: Columbia University Press, 2004), p. 2.
47 Landsberg, p. 2.
48 Landsberg, p. 15.
49 Theodor W. Adorno and Max Horkheimer, *Dialectic of Enlightenment*, trans. by John Cumming (London: Verso, 2010).
50 Gross, p. 48.
51 Grainge, p. 20.
52 Jameson, '*The Shining*'; Fredric Jameson, 'Postmodernism, or The Cultural Logic of Late Capitalism', *New Left Review*, 146 (1984), 53–92; Fredric Jameson, *Postmodernism, or The Cultural Logic of Late Capitalism* (London: Verso, 1992).
53 Jameson, '*The Shining*', p. 115.
54 Jameson, 'Postmodernism' (1984), pp. 55–57.
55 Jameson, 'Postmodernism' (1984), p. 66.
56 Jameson, *Postmodernism* (1992), p. 19.
57 Jameson, *Postmodernism* (1992), p. 25.
58 Dick Hebdige, *Hiding in the Light: On Images and Things* (London: Comedia, 1988). p. 171.
59 Theodor W. Adorno and Walter Benjamin, *The Complete Correspondence 1928–1940* (Cambridge, MA: Harvard University Press, 1999), p. 321.
60 Grainge, p. 24.
61 Grainge, p. 22.
62 Grainge, p. 43.
63 Higson, p. 126.
64 Simon Reynolds, *Retromania* (London: Faber and Faber, 2012), p. xxx.
65 Jameson, '*The Shining*', p. 115.
66 Arjun Appadurai, *Modernity at Large: Cultural Dimensions of Globalization* (Minneapolis: University of Minnesota Press, 1996), p. 78.
67 Adorno and Horkheimer, p. 126.
68 Appadurai, p. 78.
69 Nancy Martha West, *Kodak and the Lens of Nostalgia* (Charlottesville: University of Virginia Press, 2000).
70 Grainge, p. 45.
71 Grainge, pp. 44–45.
72 Daniel Fienberg, '*Fargo* Guest Bruce Campbell on the "Terror" of Playing Ronald Reagan, Shared B-Movie Background', *The Hollywood Reporter*, 9 November 2015, www.hollywoodreporter.com/live-feed/fargo-bruce-campbell-ronald-reagan-838622 [accessed 10 November 2015].
73 Sontag, *On Photography*, pp. 22–23.
74 Siegfried Kracauer, *The Mass Ornament: Weimar Essays*, trans. by Thomas Y. Levin (Cambridge, MA: Harvard University Press, 1995), p. 51; West, p. 3.

75 Niemeyer, p. 7.
76 David. R. Coon, *Look Closer: Suburban Narratives and American Values in Film and Television* (New Brunswick, NJ: Rutgers University Press, 2014), p. 39.
77 Mangan, '*Stranger Things* Review'.
78 Gil Bartholeyns, 'The Instant Past: Nostalgia and Digital Retro Photography', in *Media and Nostalgia: Yearning for the Past, Present and Future*, ed. by Katharina Niemeyer (London: Palgrave Macmillan, 2014), pp. 51–69.
79 Guiseppina Sapio, 'Homesick for Aged Home Movies: Why Do We Shoot Contemporary Family Videos in Old-Fashioned Ways', in *Media and Nostalgia: Yearning for the Past, Present and Future*, ed. by Katharina Niemeyer (London: Palgrave Macmillan, 2014), pp. 39–50.
80 Hila Shachar, 'The Ethics of Nostalgia', *Notes on Metamodernism*, 22 November 2012, www.metamodernism.com/2012/11/22/the-ethics-of-nostalgia/ [accessed 2 December 2015].
81 Hutcheon, p. 202.
82 Mark Fisher, *Ghosts of My Life: Writings on Depression, Hauntology and Lost Futures* (Alresford: O-Books, 2014), p. 25.
83 Jameson, 'Postmodernism' (1984), pp. 65–66.
84 Stefanie Coontz, *The Way We Never Were: American Families and the Nostalgia Trap* (New York: Basic Books, 2000).
85 Daniel Marcus, *Happy Days and Wonder Years: The Fifties and Sixties in Contemporary Cultural Politics* (New Brunswick, NJ; London: Rutgers University Press, 2004).
86 Owen Hatherley, *The Ministry of Nostalgia* (London: Verso, 2016).
87 Boym, *The Future of Nostalgia*, p. 342.
88 Grainge, p. 12.
89 Higson, p. 122.
90 Vera Dika, *Recycled Culture in Contemporary Art and Film: The Uses of Nostalgia* (Cambridge: Cambridge University Press, 2003); Heike Jenß, 'Cross-Temporal Explorations: Notes on Fashion and Nostalgia', 107–124; Caryl Flinn, *Strains of Utopia: Gender, Nostalgia, and Hollywood Film Music* (Princeton: Princeton University Press, 1992).

2 Memory, nostalgia, utopia and time

In the famous opening lines of T. S. Eliot's *The Waste Land* memory and desire pull in opposite directions. Memory is drawn backwards, to the warm winter, 'covering Earth in forgetful snow'; desire is drawn forward, towards the 'stirring' of something new.[1] This lyrical distillation of a contradictory temporal pull between memory and desire is compelling and intuitively resonant. But this seeming opposition is unsettled by nostalgia, for what is nostalgia if not a kind of desire steeped in memory?

As mentioned in the Introduction, there is often a perceived tension between nostalgia and utopia, which is rooted in the sense that utopia's orientation is towards the future, whereas nostalgia's orientation is towards the past. Counter to this, Ruth Levitas equates utopia with desire, a definition she arrives at in order to remind us that utopia is not in fact always directed at the future, but can just as well be directed at the past.[2] Indeed, utopia as memory of the forgotten is a significant counterpoint to the more common association of utopia with the future. This chapter will explore these competing threads in twentieth-century utopian theory, and examine the possible function of nostalgia in each.

In Chapter 3, I will argue for a more complex 'third way' in which nostalgia is understood to be neither antithetical to a future-oriented utopian drive nor aligned with a backward-looking utopian longing. Given this, this chapter may seem like something of a red herring – setting up two opposing straw-men which I will then go on to complicate. However, I believe that in order to appreciate the nuance and the original insight of this 'third way', it is necessary to understand how deeply these two binary positions have dominated utopian theory to date, not to mention more widely held conceptions of utopia and memory.

Defining utopia

Fredric Jameson distinguishes between two objects of study that have concerned students and theorists of utopia: the utopian programme – including utopian literary texts, overtly utopian political programmes and intentional communities – and the utopian impulse, which 'obscure yet omnipresent [...]

finds its way to the surface in a variety of covert expressions and practices'.[3] The first point of clarification is about my own use of the term *utopian*. In this thesis, I am concerned with the utopian impulse and the extent to which it does or does not motivate certain kinds of nostalgia; I am not concerned with literary texts or political practices that are explicitly defined as 'utopian' in the more limited sense. With this in mind, Ernst Bloch is clearly an ally, as his interest in utopia was far broader than in a particular literary genre. As he says emphatically in the introduction to *The Principle of Hope*, 'to limit the utopian to the Thomas More variety, or simply to orientate it in that direction, would be like trying to reduce electricity to the amber from which it gets its Greek name'.[4]

The second point of definitional clarification is more general. An important criterion in any definition of *utopia*, or *utopian*, is that it imagines alternatives to the way things currently are. Fredric Jameson, for example, describes utopians as those who 'offer to conceive of [...] alternate systems' and utopian politics as 'politics which aims at imagining, and sometimes even at realizing, a system radically different from this one'.[5] Zygmunt Bauman, too, equates utopianism with the human ability to think the world differently from what it currently is.[6] Although for Jameson and Bauman such alternatives are powerful and indeed necessary, utopianism need not be so positively conceived in order for the definition to stand. One might equally conceive of such imaginations as silly and pointless, while still recognising that they are alternatives. It is also worth mentioning that although 'an alternative to what is' might be a necessary definitional condition for *utopia*, it is not sufficient. Are nightmare alternatives also utopian? And death or non-existence, are they not also alternatives? However, given that the idea of utopia as an imagined alternative to what is, is a widespread basic criterion inherent in the term, the first temporal limitation on the concept of utopia arises: utopia is not the *currently present* reality. The utopian dreams towards a new configuration of reality.

So far, so simple. But if the notion of utopia as distinct from 'what is' is inherent in most understandings of the term, there are other highly charged conceptual disagreements between theorists which have important implications for whether nostalgia can ever be considered utopian. Levitas laments the lack of conceptual clarity that is typical of both theorists and historians of utopia. She suggests that 'although we may initially think we know what utopia is, when we try to define it, its boundaries blur and it dissolves before our very eyes'.[7] Her point is that when the term is used without sufficient definitional clarity, assumptions are made as to what counts as utopian and why, without these choices being properly explained. An important example of this is the intuitive and taken-for-granted belief that utopia's main concern is with the future. A further look at normative definitions of utopia will illuminate this connection.

Ideology versus utopia

One important definitional question about which there is much disagreement is whether utopia is impossible, maybe possible, or necessarily realisable. In his seminal text *Utopia*, Thomas More is often credited with naming, and thus helping to define, the tradition and practice of utopian thought (if not with inventing what surely has far older roots). The term was a humorous and deliberately ambiguous play on the Greek words 'eu-topia', meaning the good or happy place, and 'ou-topia', meaning no place. The inherence in the term of the twin ideas of perfection and impossibility has left a conceptual legacy which, as Levitas points out, 'has not been an unmitigated blessing'.[8] It has aided a colloquial understanding of the utopian as 'intrinsically impractical'[9] – dreaming the impossible dream. Today, the adjective *utopian* is commonly used in a derogatory fashion, to dismiss other people's visions as untenable, but seldom is it claimed as a descriptor of one's own hopes or aims.

As well as the common charge of impracticality, utopianism has often been accused of far worse. Marx, for example, famously dismissed the utopian socialists for their 'robe of speculative cobwebs [...] drenched with the dew of erotic sentiment'.[10] In other words, utopianism wasn't, in his view, just impractical (although it was this); it was more sinister, an escapist distraction from the actual mechanisms of social change. Marx himself, despite his stated aversion to utopianism, has of course been labelled 'utopian' by many in subsequent years, showing how deep the long-standing connection between utopianism and impossibility runs: anyone thought not to understand 'the way the world really is', we readily label utopian. Another frequent criticism of utopian thinking is that it is dangerous. Karl Popper and Friedrich Hayek, for example, view utopianism as at risk of leading to totalitarianism.[11]

Although these concerns are legitimate, particularly in light of the kinds of blueprint utopianism that characterised the twentieth century, various contemporary thinkers have a sense that there is too little utopian activity in late capitalism, that we have lost the ability to believe that society could be other than it is. Jameson explains this concern compellingly:

> What if the fact of our judgment stood as a judgment on us, rather than on the utopian speculation that we are unable to take seriously? What if our judgment were itself a measure and a symptom of our own incapacity to support such thinking, of our own repression of the principle of futurity, smothered under the realism of the reality principle and the massive weight of what is? What if it were the very psychologism, the cynical reductionism, of the reality principle itself that turned out not to be reality, but just another symptom?[12]

Ursula Le Guin, the renowned science fiction author, seconds this sentiment in a 2014 acceptance speech, which she dedicates to 'writers of the imagination, who for the last fifty years have watched the beautiful awards go to the so-called realists':

> We will be wanting the voices of writers who can see alternatives to the way we live now [...] and even imagine some real grounds for hope. We

> will need writers who can remember freedom. The realists of a larger reality. [...] We live in capitalism. Its power seems inescapable. So did the divine right of Kings![13]

The frustration evident here arises from the fact that tarring utopianism with the brush of impossibility has conveniently secured 'That Which Is', not just as a particular political-economic-historical formation that might have been or might yet be otherwise, but as the *only* realistic possibility: 'There is no alternative.'

To counter the legacy of this kind of thinking, and to reclaim the importance of utopian imagination, several theorists have built the transformative power of utopianism into their definition of utopia. For these theorists, the properly utopian is precisely that which has the capacity to change existing conditions. This notion – that utopia is not that which escapes reality, but rather that which promises to change what seems like, but is not in fact, a fixed reality is most famously articulated by Karl Mannheim. Mannheim distinguishes ideology from utopia on the basis that, although both ideology and utopia are incongruous with reality, ideology works to sustain the existing state of affairs, while utopia works to challenge and change it. An early example of ideology that Mannheim mentions is the medieval idea of paradise, which, although clearly incongruous with reality, was an integral part of maintaining social stability. Utopian ideas, by contrast, push to 'break the bonds of the existing order'.[14] In a similar vein, Ernst Bloch distinguishes between 'abstract utopia' and 'concrete utopia', with concrete (real) utopia consisting of those utopian longings that reach towards real possible futures, as opposed to dreaming that is merely wishful.[15]

There have been several pertinent criticisms of Mannheim's distinction between ideology and utopia.[16] He himself recognises that 'what in a given case is ideological and what utopian is extremely difficult' to determine.[17] This is also a difficulty with Bloch's approach, as will be explored in Chapter 3. The point here, however, is not to evaluate Mannheim's definition of utopia, but simply to observe that, against the more common sneering identification of utopia with impossibility, he defines utopia explicitly by its capacity to transform the present state of affairs.

The implications of this definitional disagreement for evaluating the utopian possibilities of nostalgia are substantive. If utopianism is considered to be impractical, impossible dreaming, then there are no limitations on its temporality. One can daydream of memories or future hopes or indeed of fantastical other realms; it does not matter which, because these dreams will not impact on reality. This is the case for those who see the purpose of utopian imagination as compensation for a less-than-perfect reality, or as harmless but naïve dreaming. On this view, nostalgia can be a utopian sentiment, but, like utopia itself, it is not very highly valued. Whether dreaming of times gone by or of alternative futures, reality will not be affected. On this view, both nostalgia and utopia are no more than escapist.

By contrast, if utopia is defined as something that might actualise and transform the status quo then the 'principle of futurity' becomes important.[18] As Levitas observes of Mannheim's distinction between ideology and utopia, it 'links utopia to the future as an anticipation and catalyst of emergent reality; while ideology is bound to the past and attempts to stabilise and preserve the status quo'.[19] On this view, nostalgia, which is necessarily oriented towards the past, is antithetical to what is properly utopian; it is aligned rather with ideology. Indeed, it is no surprise that those concerned about a lack of utopianism in late capitalism are often the same theorists who criticise contemporary expressions of nostalgia as acting to sustain the present state of affairs (see Chapter 1: Jameson, Appadurai, Fisher). The concern about a lack of utopianism in late capitalism expresses a deeper anxiety that our relationship to the future is severely impaired. Nostalgia culture, understood as a culture stuck in the past, is seen as further evidence of the triumph of ideology: despite the abundance of surface choice and possibility in this society, we are seemingly impotent to create real new futures together: '*The Future is the sign outside the No Future night club.*'[20]

Utopia: glorious past or glittering future?

Of the theorists who have created a typology of utopia, two have done so explicitly according to utopia's relationship to time. These are Zygmunt Bauman, in the final chapter of *Liquid Times: Living in an Age of Uncertainty*,[21] and Karl Mannheim, in *Ideology and Utopia*. Both typologies delineate utopia not only according to different orientations to time, but also according to the historical period in which these different experiences of time are argued to have been dominant. This attests to the interest of both theorists in understanding the historically changing nature, or expressions, of utopia; as well as in the more abstract question of the temporality of utopia.

Since the mid-1990s the sociologist and philosopher Zygmunt Bauman has been advancing a theory of 'liquid modernity'. He claims that contemporary society (leading up to and including the twenty-first century) is characterised by fluidity, uncertainty and insecurity. In *pre-modernity* social structures were fixed, roles were clear and faith in God was unshaken; in *modernity* belief in man, industry and progress overtook faith in God, and old social structures were challenged, but always with a view to creating new and better fixed structures. By contrast, in *liquid modernity* (his term for the historical phase under late capitalism, beginning towards the latter part of the twentieth century), continual change is the norm, and man (now primarily a consumer rather than producer) is struggling to adapt to the consequent insecurity of self, community and world.[22]

Bauman distinguishes three types of utopianism which he maps onto these different periods. The dominant kind of utopianism in pre-modernity is likened to *gamekeeping*, and is described as utopianism which attempts to uphold a sense of order and 'natural balance' in the world.[23] The intent here is not to improve or modernise the world, but to secure it according to either natural or divine order. In modernity a more active utopianism emerges, which Bauman refers to as *gardening*. This utopianism reflects man's aspirations to control nature and man's belief in his own agency to improve and modernise the world.[24] This gives rise to the blueprint utopias that readily spring to mind when one thinks of utopia and utopians, with their varied and competing values and proposals for how society *ought* to be. It is no surprise that Thomas More's foundational text (which is archetypal of this category) is, as Fredric Jameson points out, 'almost exactly contemporaneous with most of the innovations that seem to define modernity' (conquest of the New World, printing, Luther and modern consciousness).[25] This kind of utopianism is forward-looking, not backward-looking, and is framed by the collective narrative of progress. Finally, in liquid modernity, Bauman argues, the utopian impulse has been privatised and resembles *hunting*. The 'hunters' of today are on an insatiable and endlessly frustrating search for excitement, validation, consumer goods, and identity.[26] Although this admittedly seems a bizarre utopia, Hviid Jacobsen emphasises that for Bauman these sensation-seekers of today are nonetheless utopian, since on some level their search promises 'an ultimate and radical cure for the sorrows and pains of the human condition'.[27] Crucially, these 'utopias of the immediate' are not concerned with the worlds of yesterday or tomorrow: 'today, in liquid modern hunting utopia, previously popular utopian conceptions of [...] the "glorious past" or the "glittering future" have given way to the daily individualistic fulfilment of utopian cravings'.[28]

Karl Mannheim distinguishes four kinds of utopian mentality. He argues that due to historical changes, different mentalities have been prominent in different historical moments.[29] The first type he identifies is *Chiliasm*, which is associated particularly with Thomas Müntzer, the German theologian about whom Bloch and Engels both wrote at length. Although preoccupied with the dawn of a millennial kingdom on earth, Chiliasm was not, contrary to what one might expect, future-oriented. In fact, Mannheim identifies the chiliastic experience with 'absolute presentness', since longings which had previously attached themselves to other-worldly objectives were suddenly felt to be imminently realisable, infusing the present moment with a kind of ecstatic zeal.[30] Mannheim's second type is the *liberal-humanitarian idea*. Unlike the chiliastic mentality which was sensitive only to the 'present pregnant with meaning' (and was characteristic of the lower strata of society), the liberal-humanitarian mentality (characteristic of the middle strata of society) is preoccupied with the process of becoming.[31] The 'idea', utopia, is a goal in the future which functions to regulate existing processes and to encourage them in the right direction. His third type, *conservatism*, he describes as a counter-utopia which arises only as a means of self-defence against the attack from opposing classes.[32] Since conservatism is oriented towards the maintenance of the status quo and the power of the dominant classes, Levitas rightly questions whether, according to Mannheim's own definition, conservatism would not better be classed as an ideology. Mannheim does agree that 'conservative mentality as such has no utopia' but defends his inclusion of it as a kind of utopian thinking since it becomes conscious of its own ideals when faced with opposing theories, and then has to articulate them as an object of desire.[33] The fourth type is the *socialist-communist utopia* which, by contrast with the liberal-humanitarian idea, is not concerned with ideas but with the material conditions of oppression.

As Table 2.1 demonstrates, both Bauman's and Mannheim's typologies can be mapped onto three basic utopian categories: presentist, past-oriented and future-oriented. This is not to say that there are not distinctions within future-oriented or past-oriented utopias, but only that these temporal orientations are significant in their differences.

The inclusion of Bauman's hunters and Mannheim's chiliasts in the same category is odd, given that the communities with which they are identified could not be more different (The *hunters* are contemporary consumers in advanced Western civilisation; the *chiliasts* are Reformation-era radical revolutionaries leading the peasant wars). The only thing they have in common is their focus on the present. Yet the very idea of a present utopia goes against the claim made earlier that utopias, defined as alternatives to 'what is', are by definition not-present. And indeed, in both cases the identification as presentist is arguable. Bauman's *hunters* could be described as still future-oriented, since they do make plans and have hopes, albeit extremely short-termist and personal hopes (What new dress shall I buy tomorrow? What holiday should I go on? What drugs will make me feel good?). One could also question Bauman's description of these consumerist hopes and dreams as utopian in the first place. Mannheim's *chilaists* are presentist in that the present moment is filled with a tense and unprecedented expectation, but their energy and longing is at the same time directed to a total rupture in time as we know it. Ruth Levitas cites Paul Tillich's description of 'Kairos' to explain this as 'fulfilled time, the moment of time which is invaded by eternity'.[34] However, what both have in common is that they are 'not preoccupied [...] with optimistic hopes for the future or romantic reminiscences'.[35] This makes them outliers, as the rest of Bauman and Mannheim's types of utopianism are precisely one or other of those things: either occupied with past romantic reminiscences or with optimistic hopes for the future.

Despite important differences, what both Mannheim's *liberal-humanitarian* and *socialist-communist* utopias share is a clear focus on the future. They both entail the belief that freedom, equality and justice will prevail only in the future but the socialist-communist mentality locates this future at a much

Table 2.1 Time-related typologies of utopia

	Past-oriented	*Presentist*	*Future-oriented*
Bauman's types	*Gamekeeping*	*Hunting*	*Gardening*
Mannheim's types	*Conservatism*	*Chiliasm*	*Liberal-humanitarian idea* *Socialist-communist idea*

more specific point in time: the breakdown of capitalism. Both kinds of utopian mentality, liberal and socialist, are characteristic of the twentieth century and are aligned with Bauman's *gardening* utopianism in their belief in man's agency and power to change the way thing are. Since in this category utopia is forward-looking, both memory and nostalgia, which look towards the past, are considered regressive and as such anti-utopian.

However, memory has a central role in the other major category, past-oriented utopianism. Mannheim's *conservatism* and Bauman's *gamekeeping* are both oriented towards the past, as Mannheim remarks:

> Whereas for liberalism the future was everything and the past nothing, the conservative mode of experiencing time found the best corroboration of its sense of determinateness in discovering the significance of the past, in the discovery of time as the creator of value.[36]

These kinds of utopias embody a longing for past societies (whether remembered or imagined) which fires their desire. Another category of utopianism neither Bauman nor Mannheim mention but which deserves inclusion in the category of past-oriented utopianism is certain kinds of 'beyond this world' utopianism. Although eschatological utopias are future-oriented, other kinds of utopia are situated metaphysically beyond space and time and are concerned with that which is *outside of time*, but are in spirit backward-looking. The Garden of Eden is the prototype of this category. Often imbued with religiosity, this kind of utopianism does not intend to inspire social change but to remind us of a mystical connection and unity. It resonates with a powerful nostalgia for an imagined belonging, a sense of prelapsarian unity, before humanity's fall into the flawed, fragmented and impermanent phenomenal world. What is true for all these varieties of past-oriented utopianism is that they are aligned with memory: whether on a personal, historical or metaphysical level, utopia is something that we have to re-find or remember.

Most expressions of the utopian approximate these two extremes by exhibiting either a hopeful forward-looking expectation or a romantic backward-looking longing. To adherents of the former, nostalgia is anti-utopian; but nostalgia fuels the latter. Crucially, and contrary to Bauman's neat historicity, these backward-looking utopias are no less at home in modernity than forward-looking utopias. Just as Boym observes that nostalgia and progress, like Jekyll and Hyde, are both coeval with modernity, these competing temporal tendencies (backward-looking or forward-looking) are likewise in tension within modern utopian thought. Richard Tarnas has argued that this reflects the competing tendencies of Enlightenment and Romanticism at play in modernity, what he calls two basic myths of historical self-understanding: the myth of Progress and the myth of the Fall:

> These two historical paradigms appear today in many variations, combinations and compromise formations. They underlie and influence discussions

> of the environmental crisis, globalization, multiculturalism, fundamentalism, feminism and patriarchy, evolution and history. One might say that these opposing myths constitute the underlying argument of our time: Whither humanity? Upward or downward? How are we to view Western civilization, the Western intellectual tradition, its canon of great works? How are we to view modern science, modern rationality, modernity itself? How are we to view 'man'? Is history ultimately a narrative of progress or of tragedy?[37]

What is important, for present purposes, is that this binary leaves limited options for understanding nostalgia. It is a stark choice between those who view the past as 'the good ol' days', that is, as a legitimate object of desire to which they wish to return, and those who insist on the transformative necessity of imagining alternative futures. If you value the future, it seems, there is no place for nostalgia.

Marcuse and Bloch: psychoanalysis, remembrance and the utopian impulse

As already mentioned, my own concern is with the utopian impulse, rather than with the more overt utopianism that Jameson calls the utopian programme. Interestingly, though, the same basic temporal orientations that have predominantly governed the utopian programme – forward-looking or backward-looking – are also often identifiable in theorisations of a utopian impulse. In the Introduction I mentioned that among theorists belonging to, or associated with, the Frankfurt School, one can find some of the most nuanced and powerful theorisations of the utopian impulse. In Chapter 3 I will explore aspects of Ernst Bloch's thought (as well as of Herbert Marcuse's and Walter Benjamin's) that offer ways of thinking about nostalgia that go beyond the bad/good, anti-utopian/utopian dichotomies, that persist in most histories or theories of utopia. However, I first want to point out that *even these* highly nuanced thinkers sometimes reproduce the two prominent temporal narratives of utopia. This suggests the importance of those parts of their work that in some way unsettle this dichotomy. My approach rests on the belief that an author's body of work is not always of a piece: one can find strains, hints and shadows of different philosophical positions in a thinker's oeuvre. Moreover, where different philosophical legacies (and emergences) compete in a body of work, this is not always conscious, and seldom clearly articulated by the thinker in question. This is where the work of interpretation comes in.

I will now consider Bloch and Marcuse, showing how these two theorists come at the question of a utopian impulse from opposing temporal directions. Jameson touches on this in the following short passage which he does not elaborate on, but which I take as an interesting starting point for considering the role that the past and the future play for these two theorists of the utopian impulse:

> [...] Bloch's central concern as a philosopher, namely the blindness of all traditional philosophy to the future and its unique dimensions, and the denunciation of philosophies and ideologies, like Platonic anamnesis, stubbornly fixated on the past, on childhood and origins. It is a polemical commitment he shares with existential philosophers in particular, and perhaps more with Sartre, for whom the future is praxis and the project, than with Heidegger, for whom the future is the promise of mortality and authentic death; and it separates him decisively from Marcuse, whose Utopian system drew significantly, not merely on Plato, but fully as much on Proust (and Freud), to make a fundamental point about the memory of happiness and the traces of gratification that survive on into a fallen present and provide it with a 'standing reserve' of political and personal energy.[38]

Janet Stewart points out that of the Frankfurt School theorists Marcuse 'most obviously employs a theory of remembrance directly derived from Psychoanalysis to approach utopia.'[39] Ernst Bloch, I will show, most clearly critiques psychoanalysis for its emphasis on remembrance, drawing attention instead to the forward-dawning possibilities for the psychology of a utopian impulse.

Marcuse and remembrance

Martin Jay observes that many critics have noted the importance of memory for Marcuse, and he outlines four different sources of Marcuse's fascination with memory.[40] First, his early philosophical apprenticeship under Heidegger. Although he later came to reject Heidegger's notion of Being, he retained his teacher's sense that something very important had been lost in the modern world.[41] Second, his adherence to critical Marxism. There was a sense among his key contemporaries that humanity had been in some way mystified under capitalism, that objects had been made thing-like and that something of them had thereby been forgotten.[42] This is most famously articulated by Adorno's phrase 'all reification is forgetting', and from this Marcuse came to the insight that remembrance could be a source of resistance to reification.[43] Third, his special concern for aesthetics. Although Marcuse rarely acknowledged a debt to the Romantic tradition, Jay suggests that he was always drawn to the Romantics and that their influence is clear.[44]

The fourth and most significant source of Marcuse's fascination with memory is his radical appropriation of psychology, which leads most obviously to his theory of remembrance.[45] Marcuse takes as his starting point Freud's claim in *Civilization and Its Discontents* that suppression of Eros (the life instinct) is necessary since the *pleasure principle* and the *reality principle* are in conflict with one another, and that civilisation is built on this necessary repression.[46] However, he argues contra Freud that since this conflict between the pleasure principle and the reality principle is based on a condition of scarcity, it is historically contingent. In particular, technological progress has made possible the virtual abolition of scarcity, so the repression of the pleasure principle that may once have been necessary in order for society to work well is no longer necessary. He thinks that some degree of repression of the pleasure principle might be necessary in a functioning society, but that contemporary society is characterised by more repression than is needed: there is, in other words, an element of surplus repression. These 'additional controls over and above those indispensable for civilized human association' help maintain the domination of the ruling groups and those in power.[47]

Marcuse also claims that the specific form of the reality principle that is operative in advanced industrial civilisation is the *performance principle*, a principle which keeps people working harder than is necessary.[48] It even manages to infiltrate the few hours of free time by the creation of false needs: most often the individual 'desires what he is supposed to desire'.[49] As Marcuse bitingly observes, 'The people recognize themselves in their commodities; they find their soul in their automobile, hi-fi set, split-level home, kitchen equipment.'[50] It is worth noting that Marcuse, an emigrant to America, was witnessing a totally new post-war social environment which combined a greater abundance and availability of consumer goods with more sophisticated forms of social control such as, for example, more effective advertising methods. Marcuse was grappling with the felt complexities of that era, which continue to exert influence on contemporary American society and culture today. It was in this alienating context that Marcuse's urgent utopianism emerged, as Jameson explains:

> [...] by the philosopher, in the exile of that immense housing development which is the state of California, remembering, reawakening, reinventing – from the rows of products in the supermarkets, from the roar of the freeways and the ominous shape of the helmets of traffic policemen, from the incessant overhead traffic of the fleets of military transport planes, and as if it were from beyond them in the future – the almost extinct form of the Utopian idea.[51]

'Remembering, reawakening, reinventing': the role of memory in utopia is central for Marcuse. As discussed, he saw the *performance principle* as responsible for unnecessary repression of Eros, and it is in the 'aesthetic-erotic' dimension of Eros that, for Marcuse, utopian hope was contained. Crucially, he sees Eros as something that *was once known* but has since been repressed. This idea derives from an unorthodox reading of Freud. He takes the notion of repression from Freud, but whereas Freud emphasises the repression of painful or traumatic past experiences, Marcuse argues that early experiences of gratification (or Eros) are also often repressed, in order to enable survival in a society that demands *performance* over *pleasure*.[52] He argues that 'the replacement of the pleasure principle by the reality principle is the great traumatic event in the development of man', and that 'the psychoanalytic liberation of memory explodes the rationality of the repressed individual. As cognition gives way to re-cognition, the forbidden images and impulses of childhood begin to tell the truth than reason denies'.[53]

In Marcuse's interpretation of Freud this recollection of previous gratification is not only powerful for the individual, but it also functions at the level of civilisation. He posits an archaic pre-history before surplus repression took hold of 'reality', in which the pleasure principle was supreme. The actual status of this so-called pre-history is not so much historical as symbolic. For Marcuse, the memory of 'the full force of the pleasure principle' (for the most part a mythic symbolic memory rather than an actual historical memory) survives in the unconscious: 'The *return of the repressed* makes up the tabooed and subterranean history of civilization.'[54] And it is this memory of gratification – memory of the full possibilities and power of Eros – that fires the utopian imagination. The kind of society Marcuse goes on to describe in *Eros and Civilization* as possible and desirable encapsulates 'receptivity, non-violence, tenderness' – all the characteristics pertaining to the long-forgotten, re-remembered Eros.[55]

This notion of an archaic pre-history from which one can draw utopian energies has been criticised by Martin Jay as insufficiently critical. Jay argues that 'Marcuse was obviously introducing here a myth of original wholeness, of perfect presence, of the "re-membering" of what had been dismembered' which he sees as smuggling in an a priori philosophical anthropology into critical theory.[56] Jameson, too, points out that the theoretical foundation which for Marcuse is the groundwork for the very possibility of utopian thinking, 'takes the form of a profound and almost Platonic valorisation of memory, anamnesis, in human existence'.[57]

What is interesting for present purposes is that even Marcuse, the utopian philosopher who is best known for his influence on 1960s counterculture (the 1960s – that time about which Sontag wrote that 'there was so little nostalgia'), whose ideas were taken up by young people espousing the then radical notion of free love and a rejection of consumerism, had, at the core of his philosophy, a mythic symbolic nostalgia for an original wholeness (a land of Eros) that has been forgotten. Marcuse's utopianism, although hopeful for a better future, derives its heart from remembrance of a forgotten past.

There is another way, too, in which Marcuse might be accused of nostalgia, and that is in relation to his theory of the utopian potential of art. Marcuse sees art as having a transformative role to play in society, one that needs defending from a traditional Marxist perspective which discounts art's potential role. For him, art, fantasy and imagination are all central to sustaining and realising the vision of an alternative way of being.[58] However, he believes that at the time of his writing art had lost some of the subversive power it once had: 'The progress of technological rationality is liquidating the oppositional and transcending elements in the "higher culture".'[59]

This is a historical nostalgia. Although Marcuse recognises that the pre-technological era was profoundly unequal and hard and that high culture was confined to privileged minorities in a world otherwise characterised by the misery of labour, he still thinks that high culture in that era was free in a way that contemporary culture rarely is. Feudal culture, he suggests, was in itself free from the motives of business and profit and so had not yet thoroughly organised man and nature into instrumentalities. It spoke, instead, with 'the rhythm of those who wander or ride in carriages, who have the time and the pleasure to think, contemplate, feel and narrate'.[60] Even as the bourgeois order emerged, Marcuse argues that for a while high culture contained within it elements of estrangement from and aesthetic incompatibility with the emerging industrial society. Admittedly, this 'high culture' was allowed to continue on the assumption of a separation of spheres – that is, it did not in real terms disturb the order of business. Nonetheless, it contained within it images ('and the dimension which they reveal: Soul and Spirit and Heart') that were genuinely distinct and distant from the daily realities and motives of business, and in that distance lay the grounds of their truth.[61]

In advanced industrial society, by contrast, literature and art continue but their antagonistic contents have been assimilated and absorbed in a kind of cultural reconciliation. Culture has found a way to absorb powerful and antagonistic imagery, to rob it of its power to estrange, condemning potentially powerful works of art to 'the fate of being absorbed by what they refute'.[62] As he comments damningly, 'the cultural centre is becoming a fitting part of the shopping centre'.[63]

Crucially, Marcuse diagnoses an increasingly widespread bureaucratic logic which invalidates the radical otherness and inexplicability (and therein the truth) of previously powerful artistic and literary works by subjecting them to a rationale of explanation and reduction:

> The psychiatrist takes care of the Don Juans, Romeos, Hamlets, Fausts, as he takes care of Oedipus – he cures them. [...] The soul contains few secrets and longings which cannot be sensibly discussed, analysed, and polled.[64]

To be clear, Marcuse is not idealising earlier stages in society as themselves utopias, as is the case with Mannheim's *conservatives* or Bauman's *gamekeepers*. What he thinks is lost is not a more perfect society, but a society which was less totally dominated by the logic of capitalism and instrumental reason, and thereby more open to the utopian dimension; a society in which repression could still be seen more clearly for what it was, and in which spaces of resistance still existed – most notably in art but also in sexuality – which allowed at least for the conceptual possibility of negation. Nonetheless, the belief that earlier societies were at least relatively freer, since the artistic space was less fully incorporated into capitalist logic, is another feature of Marcuse's search for utopia that could be seen as coloured by nostalgia.

Bloch and the not yet

In addition to their biographical similarities, mentioned in the Introduction, Bloch and Marcuse share a number of central philosophical tenets. Ontologically, for both, the utopian ideal involves a union of subject and object.

Jameson describes Marcuse's remembrance of 'a time that precedes the very separation of the subject from its object' and Bloch's anticipation of that same union: 'The marvellous is a lightning bolt from subject and object alike, in the light of which there no longer exists anything alienated, and in which subject and object have simultaneously ceased to be separated from each other.'[65] For both, too, the role of art in articulating alternatives is central. They also share the belief that bourgeois idealism and its products (art, but also religion) are not simply ideological (and hence demand only destruction or exposition) but also contain shards of hope – of truth.[66]

However, they approach the question of remembrance, and the relationship between memory and utopia, from very different directions. Both Bloch and Marcuse go to some lengths to understand the psychology of the utopian impulse, and introduce their own ideas by way of an extended analysis of Freud (Marcuse primarily in *Eros and Civilization*, and Bloch primarily in Book One of *The Principle of Hope*). But while Marcuse sees himself as largely loyal to Freud (although others have disagreed, seeing his interpretations of Freud as a distortion of the original material), Bloch is overtly critical of Freud.

Bloch's initial criticisms of Freud, with which he opens his discussion, are familiar and reflect a classically Marxist orientation. He questions the nature of a basic human drive, whether Freud's 'libido' or Adler's 'will to power', is a bourgeois misrepresentation arising out of the 'class-based limitation of psychoanalytic research'.[67] Bloch argues that man has multiple drives, and that different ones are prevalent in different socio-historical moments. Freud's mistake, he thinks, is to naturalise one: 'The historical location of "the human creature" as characterised by Freud, can [...] be precisely determined: this libido-man lives – together with his dreamed wish-fulfilments – in the bourgeois world a few decades before and a few decades after 1900.'[68] Indeed, if any one factor deserves the status of a basic drive, according to Bloch, it is hunger, 'the drive that is always left out of psychoanalytic theory'.[69] Although Bloch respects Freud's contribution to explaining an important aspect of the human psyche – that is, that 'forgotten wounds and disappointments continue to smart; they have disappeared from the conscious ego, but not from the psyche. From them derive seemingly unfounded sensitivity, over-reaction, compulsive neurotic activity' – he rejects the notion that *all* anxiety can be accounted for in terms of repression.[70] He asks, 'Are there not also Objects, circumstances, which are menacing enough in an *object-based* way?' He mentions hunger, subsistence worries, economic despair and, for example, 'the anxiety caused by fascism [...] which hardly needed the pretext of infantile trauma in order to be delivered into the world'.[71]

This whole line of criticism, although important, is not new. More interesting, particularly in light of its centrality for Marcuse, is Bloch's attack on Freud's theory of repression. Bloch's central criticism of Freud is, simply put, that *'there is nothing new in the Freudian unconscious'*. Even the psychoanalytic method of making the unconscious conscious, 'only clarifies What Has Been'. All that is revealed in uncovering repressed memory is 'the reeled-up

Long Ago'.[72] A Freudian would no doubt argue that the psychoanalytic method does not neglect the new, that it focuses on releasing the psyche from that which hinders it, just as, according to Terry Eagleton, Marxism aims to 'resolve the contradictions in the present which prevent a better future from coming about'.[73] However, Freudian theory describes human development as well as articulating normative therapeutic practice, and Bloch is right to draw attention to the fact that little in-depth theoretical analysis has been devoted to understanding the part of the human psyche that is able to look forward, even if many psychoanalytic practitioners allow for and encourage that possibility.[74] Although Bloch accepts Freud's mastery in bringing to light the reality of repressed psychic material, he laments that 'there is as yet no psychology of the other side, of forward dawning'.[75]

By contrast, Bloch's central concern is with that which is 'Not Yet'. Bloch is said to have discovered the concept of the *Noch Nicht* at the age of twenty-two and was so fascinated by it that it became a cornerstone of his philosophy up until his death in 1977.[76] For him, the human psyche could not be understood solely in terms of 'sublimations, infantilisms and archaisms' but also needed an account of the forward-looking creative impulse that has the capacity both to dream up, and dream into being, possibilities which have not been before.[77] Crucially, since we cannot clearly think that which has not been thought before – we do not yet have the language to do so – it is within the unconscious that the possibility of the Not Yet first arises. So the unconscious contains two elements: the Freudian No Longer Conscious, consisting of repressed material from earlier wounds; and Bloch's own category, the Not Yet Conscious, 'a creative source of material on the verge of coming into consciousness'.[78] The Not Yet Conscious is 'the psychological birthplace of the New'.[79] The mode of awareness that pertains to the Not Yet Conscious is a pre-appearance that points toward the possibility of the Not Yet Become. In other words, it is in the unconscious that real possibilities for a genuinely new future first emerge.

Bloch is a poet as well as a philosopher, and the language he uses in relation to hoping and dreaming is steeped in semi-consciousness, reminiscent of Yeats's 'The blue and the dim and the dark cloths/of night and light and half-light'.[80] For example, Bloch writes that 'the day too has twilight edges', about 'the blue hour between daylight and darkness' and that 'the overlap of the black and blue hours happened again every time both were proud not to be day in the sense of superficial clarity, mere superficial connection'.[81] It is in this Not Yet Conscious 'bright-dark mood' that resistance both to the instrumental logic of the 'daylight' of the already defined world and to the regressive lure of the Freudian night-world is first felt: on this edge, a restless sense of the possibility of the New. Here, not in lucid thought, is where the seeds of hope are born.[82]

Bloch's critique of Freud echoes his broader attack on 'the whole of Judeo-Christian philosophy, from Philo and Augustine to Hegel' that 'the Ultimum relates exclusively to a Primum and not to a Novum'.[83] As Jameson observes in his essay on Bloch in *Marxism and Form*, 'what strikes the random observer of the history of philosophy is the lack of attention given the future as such, so that there was something essentially frivolous in a consideration of that which does not yet exist'.[84] Bloch's entire philosophy of hope can be read on one level as a rallying cry against this neglect of the future as a category in the history of philosophy.

For Bloch, against Heidegger's notion of Dasein, nobody has ever yet really *been*, nobody has ever yet really lived.[85] The promise of a fully actualised humanity looms as a possible horizon towards which the utopian tendency reaches out.

To Bloch, it is a fundamental mistake present in much philosophy to read memory – either psychologically in early childhood or ontologically in archaic anamnesis – as the forgotten source of perfection, the remembrance of which informs utopia. Instead, he is drawn to 'the *topos of the unknown within the future*, where alone we occur, where alone, novel and profound, the function of hope also flashes, without the bleak reprise of some anamnesis'.[86] His rallying cry is thus not only for recognition of 'the future' as the most important hitherto overlooked category but equally against what he sees as the sneaky hold of theories of anamnesis. At least in Freud, he suggests, 'the sick person was only reminded of the unconscious so that he could free himself from it' whereas in Jung 'libido becomes archaic; blood and soil, Neanderthal man and Tertiary period leap out simultaneously to confront us'.[87] Romanticism is described as a 'nostalgic grave'.[88] For Bloch, this whole line of thinking, the adherence to which is clear in Marcuse, is bogus. In Bloch's view, there is no '"original" drive, nor a "primal man" or even an "old Adam"'.[89]

This difference between Bloch and Marcuse is apparent in their view of the role of art. For both, imagination and fantasy come to the fore in art and therein lies the power of the artwork to envision genuine alternatives to the world as it currently is. Art (including, of course, literature and music) is central in bringing forth a really possible utopian world. However, for Marcuse, what returns in art is the repressed: the archaic memory of both gratification and Eros. It is this archaic knowledge, on some deep level, which informs future revolutionary aesthetic-erotic practice. For Bloch, too, it is what is suppressed by society that motivates the artwork, but for him this is not an archaic memory, but the creative aspect of the unconscious – the Not Yet Conscious. It is this creative, spontaneous, utopian forward-dawning that emerges into the light in art:

> Therefore great works of art can dispense least of all with the creative touch of poetic anticipation – not the concealment or repression but the pre-semblance of what, objectively, is still latent in the world.[90]

A false binary?

In this chapter I have shown that two characterisations of the relationship between memory and utopia have dominated twentieth-century utopian theory and the instances of utopian expression on which the theorists reflect: memory as the forgotten source of utopia (the 'glorious past') and memory as antithetical to utopia, which reaches out towards the time to come (the 'glittering future'). In histories of the utopian programme, this binary is clear, as was demonstrated via an analysis of Bauman's and Mannheim's typologies of utopianism. The implication of this binarism for my own question – whether nostalgia might ever be considered utopian – seems to be that nostalgia might be considered utopian only if one conceives of the past as more perfect than the present, and desires somehow to recreate it. If one values the future, by contrast, nostalgia is quite the opposite of utopian: it is ideological, and should be rejected out of hand.

A first look at Bloch's and Marcuse's theories of the utopian impulse highlights just how deep-rooted this legacy is. Marcuse, as I have explained and as several commentators have pointed out, has a deep nostalgia at the heart of his utopian philosophy: primarily for a mythical remembered moment before Eros was repressed, and secondarily for earlier historical societies as having been at least relatively freer. Bloch, by contrast, rejects the sneaky hold that theories of anamnesis have on most philosophy, and instead roots firmly for the future, the new, and the emergence of hitherto unknown possibilities, as the core of his utopianism. In so doing, he makes clear his disregard for most kinds of philosophical remembrance, nostalgia included. This seems to confirm the stark choice between nostalgia and belief in the possibility of the future.

However, although my presentation of Marcuse and Bloch here draws out important, and real, elements of their thinking, it does not do justice to the complexity of either's theories on the subject of memory and hope. I thought it necessary to concede that even the most nuanced thinkers of a utopian impulse, indeed the thinkers who have offered the richest theorisations of nostalgia and hope, are not beyond the age-old binaries of philosophies of nostalgia and hope – beyond, that is, the question, as Tarnas puts it: 'Whither humanity? Upward or downward?'; Marcuse at times seems to draw uncritically on a kind of Platonic anamnesis as the source of utopia, and Bloch, at times, in his keenness to resist this, seems too dismissive of the role that memory might play. But in Chapter 3 I will explore different aspects of their thought, especially the parts that unsettle this dichotomy. In so doing, I will highlight an important 'third way' between the two dominant narratives presented in this chapter.

Notes

1 T.S. Eliot, 'The Waste Land', in *The Waste Land and Other Poems* (London: Faber and Faber, 1972), pp. 21–46 (p. 23, II. 4–6).
2 Levitas, pp. 220–222.
3 Jameson, *Archaeologies of the Future*, p. 5.
4 Bloch, *The Principle of Hope*, I, p. 15.
5 Jameson, *Archaeologies of the Future*, p. xii.

6 Michael Hviid Jacobsen, 'Liquid Modern "Utopia" – Zygmunt Bauman on the Transformation of Utopia', in *Utopia: Social Theory and The Future*, ed. by Michael Hviid Jacobsen and Keith Tester (Farnham: Ashgate, 2012) pp. 69–96 (p. 73).
7 Levitas, p. 2.
8 Levitas, p. 3.
9 Levitas, p. 3.
10 Karl Marx and Friedrich Engels, *The Communist Manifesto*, ed. and trans. by L. M. Findlay (Peterborough, ON: Broadview Press, 2004), p. 88.
11 Levitas, p. 3.
12 Jameson, *Marxism and Form*, p. 90.
13 Ursula Le Guin, Acceptance Speech for the National Book Foundation's Medal for Distinguished Contribution to American Letters at the 65th National Book Awards, online video recording, YouTube, 19 November 2014, www.youtube.com/watch?v=Et9Nf-rsALk&app=desktop [accessed 12 November 2014].
14 Karl Mannheim, *Ideology and Utopia: An Introduction to the Sociology of Knowledge* (London: Routledge, 1960), p. 173.
15 Although this distinction is clear in Bloch, his views of abstract utopia are not entirely dismissive, as will become clear.
16 In *Lectures on Ideology and Utopia*, ed. by George H. Taylor (New York: Columbia University Press, 1986) Paul Ricoeur criticises Mannheim's assumption that reality is easy enough to identify sociologically, arguing that social reality always involves human beings and their thoughts and is therefore hard to distinguish from ideology. He also criticises the allocation of realisability as a criterion of utopia in that it doesn't allow us to derive its pathology as wishful thinking (p. 176). Another criticism is levelled by Levitas (p. 90), who draws on Raymond Williams to point out that the dominant culture is not homogeneous ('reality') but contains alternatives within it.
17 Mannheim, p. 176.
18 Jameson, *Marxism and Form*, p. 90.
19 Levitas, p. 87.
20 Bloch, *The Principle of Hope*, I, p. 5.
21 Zygmunt Bauman, *Liquid Times: Living in an Age of Uncertainty* (Cambridge: Polity Press, 2007).
22 Zygmunt Bauman, *Liquid Modernity* (Cambridge: Polity Press, 2000). See 'Foreword: On Being Light and Liquid' for a more complete discussion.
23 Bauman, *Liquid Times*, pp. 98–99.
24 Bauman, *Liquid Times*, p. 99.
25 Jameson, *Archaeologies of the Future*, p. 1.
26 Jacobsen, p. 76.
27 Jacobsen, p. 76.
28 Jacobsen, p. 76.
29 Mannheim (pp. 189–190) points out that 'they are simply methodological devices. No individual mind, as it actually existed, ever corresponded completely to the types' but still suggests that 'each individual mind in its concreteness, however (despite all mixtures), tends to be organised in general along the structural lines of one of these historically changing types'.
30 Mannheim, pp. 190–191.
31 Mannheim, p. 202.
32 Mannheim, p. 207.
33 Mannheim, p. 206.
34 Levitas, p. 82.

35 Mannheim, p. 195.
36 Mannheim, p. 211.
37 Richard Tarnas, *Cosmos and Psyche* (New York: Viking Penguin, 2006), p. 13.
38 Jameson, *Archaeologies of the Future*, p. 7.
39 Janet Stewart, p. 34.
40 Martin Jay, 'Anamnestic Totalization: Reflections on Marcuse's Theory of Remembrance', *Theory and Society*, 11.1 (1982), 1–15 (p. 1).
41 Jay, p. 2.
42 Jay, p. 5.
43 Adorno and Benjamin, p. 321.
44 Jay, p. 6.
45 Jay, p. 7.
46 Herbert Marcuse, *Eros and Civilization*, pp. 3–5.
47 Marcuse, *Eros and Civilization*, p. 37.
48 Marcuse, *Eros and Civilization*, p. 37.
49 Marcuse, *Eros and Civilization*, p. 46.
50 Herbert Marcuse, *One-Dimensional Man* (London; Routledge, 2002), p. 11.
51 Jameson, *Marxism and Form*, p. 116.
52 Jay, p. 8.
53 Marcuse, *Eros and Civilization*, p. 15, p. 19.
54 Marcuse, *Eros and Civilization*, p. 16.
55 Levitas, p. 159.
56 Jay, p. 10.
57 Jameson, *Eros and Civilization*, p. 112.
58 Levitas, p. 164.
59 Marcuse, *One-Dimensional Man*, p. 59.
60 Marcuse, *One-Dimensional Man*, p. 63.
61 Marcuse, *One-Dimensional Man*, p. 64.
62 Marcuse, *One-Dimensional Man*, p. 74.
63 Marcuse, *One-Dimensional Man*, p. 68.
64 Marcuse, *One-Dimensional Man*, p. 74.
65 Jameson, *Marxism and Form*, pp. 113, 141.
66 Levitas, p. 170.
67 Bloch, *The Principle of Hope*, I, p. 66.
68 Bloch, *The Principle of Hope*, I, p. 68.
69 Bloch, *The Principle of Hope*, I, p. 64.
70 Bloch, *The Principle of Hope*, I, pp. 54–55.
71 Bloch, *The Principle of Hope*, I, p. 84.
72 Bloch, *The Principle of Hope*, I, p. 56.
73 Terry Eagleton, *Why Marx Was Right* (Yale: Yale University Press, 2011), pp. 68–69.
74 Existential Psychology is an exception to this, but as a movement it also distinguishes itself clearly from Freud.
75 Bloch, *The Principle of Hope*, I, p. 116.
76 Caroline Edwards, 'Uncovering the "Gold-Bearing Rubble": Ernst Bloch's Literary Criticism', in *Utopianism, Modernism, and Literature in the Twentieth Century*, ed. by Alice Reeve-Tucker and Nathan Waddell (London: Palgrave Macmillan, 2013), pp. 182–203 (p. 187).
77 Bloch, *The Principle of Hope*, I, p. 86.
78 Levitas, p. 101.
79 Bloch, *The Principle of Hope*, I, p. 116.

80 William Butler Yeats, 'He Wishes for the Cloths of Heaven', *Online-literature.com*, www.online-literature.com/frost/776/ [accessed 1 May 2014].
81 Bloch, *The Principle of Hope*, I, p. 101; Bloch, *The Principle of Hope*, I, p. 86; Bloch, *The Principle of Hope*, I, p. 86.
82 Bloch, *The Principle of Hope*, I, p. 107.
83 Bloch, *The Principle of Hope*, I, p. 203.
84 Jameson, *Marxism and Form*, p. 125.
85 Jameson, *Marxism and Form*, p. 136.
86 Ernst Bloch, *The Spirit of Utopia*, trans. by Anthony A. Nassar (Stanford: Stanford University Press, 2000), p. 201.
87 Bloch, *The Principle of Hope*, I, p. 59.
88 Bloch, *The Principle of Hope*, I, p. 160.
89 Bloch, *The Principle of Hope*, I, p. 67.
90 Ernst Bloch, *A Philosophy of the Future*, trans. by John Cumming (New York: Herder and Herder, 1970), p. 96.

3 Nostalgia as 'gold-bearing rubble'

In looking for a way of reading contemporary nostalgia culture that is more hopeful than Fredric Jameson's influential critique of the nostalgia mode, Ernst Bloch might at first seem an unlikely resource. As discussed in Chapter 2, he is most renowned for his committed and continuing efforts to expound the 'future' and the 'new' as philosophical categories, which he believed had been previously neglected by nearly all major systems of thought. More than this, he launched a full frontal attack on the doctrine of memory as the 'absolute inversion' of hope.[1] His most in-depth and pointed critique is of Freud, but he concerns himself also with rooting out and demolishing the pernicious influence of all theories of *anamnesis*, including the 'Romantic reactionary distortions' of Carl Jung and those of the 'sentimental penis-poets like D.H. Lawrence' and the 'complete Tarzan philosophers like Ludwig Klages'.[2] Utopianism nostalgic for a past 'Golden Age' is met with a similarly damning appraisal.[3] To locate perfection in the past is, for Bloch, a fundamental error.

Despite the overall thrust of Bloch's orientation towards the future, however, he is not dismissive of the past tout court. On the contrary, he is highly critical of many of his fellow Marxists for relying too heavily on an oversimplified and inadequate theory of progress. It is in this context that he demonstrates a nuanced and powerful understanding of memory, nostalgia and utopia, which is very relevant to this work. My use of Bloch relies both on his approach to ideology critique in general, and on how he applies his ideology critique in relation to non-contemporaneity.

Bloch's ideology critique

Rather than seeing ideology and utopia as opposites, as Mannheim does (see Chapter 2), Bloch thinks that ideological elements often permeate (and distort) utopia and likewise utopian elements permeate ideology. He disagrees with accounts which simply denounce all ideology as false consciousness, or see ideology exclusively as an instrument of 'mystification, error and domination'.[4] Rather, as Douglas Kellner explains, Bloch sees in ideology *both* deceptive-illusory elements and emancipatory-utopian elements.[5] For Bloch, the primary site of ideology is the cultural superstructure, but the superstructure cannot be reduced entirely to ideology. This has parallels with Raymond Williams' comment that 'no mode of production, and therefore no dominant society or order of society, and therefore no dominant culture, in reality exhausts human practice, human energy, human intention'.[6] Throughout Bloch's analysis of daydreams, myths, popular culture, literature, political praxis, religion and more in *The Principle of Hope* and elsewhere, as Kaufman puts it, he 'presents himself as a redemptive reader of what used to be called the superstructure'.[7] He urges us to pay attention to the potentially progressive or utopian content in 'phenomena frequently denounced and dismissed as mere ideology'.[8]

This is an unorthodox approach to cultural criticism for someone coming from a Marxist perspective, although some would argue that it is closer in spirit to what Marx himself intended. Typically, Marxist cultural critics have adopted a method of exposition in which the goal of ideology critique is to expose false consciousness in order to allow an altogether different consciousness to flourish. For Bloch, by contrast, ideology critique does not require demolition, but hermeneutics.[9] As Kellner puts it, 'one of the tasks of radical cultural criticism is to specify utopian, critical, subversive, or oppositional meanings, even within the texts of so-called mass culture'.[10] Bloch's hermeneutic thus involves tracing hidden moments of possible hope within the ordinary: 'glowing emblems' where the forward-dawning of the utopian impulse shines through.[11] Identifying utopian content within ideology requires looking out for traces and signs which might not at first announce themselves:

> the imperceptible tending of all things towards Utopia, of the future which stirs at its convulsive but microscopic work within the smallest cells of the vast universe itself, make themselves known to us as *Spuren* in the world both without and within: traces, spoor, marks and signs.[12]

Already, this approach offers us a different way of looking at contemporary cultural nostalgia to the way that it has typically been approached. As we saw in Chapter 1, Jameson's critique of the 'nostalgia mode' identifies it entirely as a depthless reflection, and ideological lackey, of late capitalism. Likewise, Appadurai's analysis of 'ersatz' or 'armchair' nostalgia is devoted entirely to identifying it with marketed nostalgia, nostalgia for sale. Lipovestky is similarly damning of the nostalgia industry as 'tourism of memory', which transforms real memory into entertainment and spectacle.[13] Contra the total denunciation of contemporary forms of nostalgia by these critics, each of whom seeks to expose nostalgia culture as ahistorical, depthless, consumerist, etc., my readings of nostalgia texts in Part II respond to Bloch's call for a nuanced sifting through mainstream culture as through 'gold-bearing rubble', and seek to identify the authentic utopian core at work in these nostalgic longings.[14]

Such a method, Žižek reminds us, is particularly important today:

> In our 'postmodern' cynical constellation, he [Bloch] reminds us that denunciation of ideology is not enough: every ideology, even the most horrifying Nazism, exploits and relies on authentic dreams, and to combat false liberation one should learn to discern in it the authentic utopian core.[15]

However, just as Mannheim admitted that in practice distinguishing ideology from utopia was no easy task, so distinguishing the utopian-emancipatory from the deceptive-illusory within mass cultural texts, as a Bloch-inspired approach requires, is in practice harder than in theory. Searching through 'gold-bearing rubble' for the authentic-utopian raises the question: How does one distinguish between the emancipatory and the illusory? How does one identify the utopian impulse at work? This is not, of course, a straightforward process of checking off certain criteria. Instead, one has to rely to an extent on persuasive critical analysis. Since elaborating the meaning of utopian consciousness was Bloch's core preoccupation throughout a lifetime of scholarship, a neat one-line definition of his conception of the utopian must elude us. However, if we look at a few of the descriptions that surface repeatedly in his work, an idea of what he means by *utopian* will start to emerge. This might then serve as a guide in the present search for utopia within contemporary nostalgia culture. In what follows, I offer four distinct but related conceptions of the utopian that Bloch offers which, taken together, paint a picture of what aspects of nostalgia culture I might be able to identify as having utopian-emancipatory potential.

First, the most uncomplicatedly Marxist description that Bloch gives of the utopian within ideology, the '*genuine nebulae* (which still have to give birth to a star)', is of those elements that are 'hostile to capitalism, homeless within it'.[16] This is fairly central, and remains a solid basic criterion, a reminder of the tradition within which Bloch was writing and to which this book remains broadly committed. If utopianism, on nearly all accounts, involves the imagination of alternatives to what is, 'what is', on a systemic level today, is capitalism. To describe something as utopian or as motivated by the utopian impulse requires, therefore, that it reaches towards possibilities that are felt to be lacking in the currently dominant culture, rather than merely reflecting the hegemony.

Second, Bloch insists:

> Deprivation lacks food and, in the centre, something higher as well. [...] This habitually, and ultimately 'spiritually' missed element thus likewise contradicts the Now, just as powerfully as the lacking food.[17]

This statement reflects Bloch's broad understanding of the nature of contemporary alienation, and of the strong influence of Christian mysticism on his thinking. This more mystical conception of the utopian impulse as seeking spiritual fulfilment begs the question: What concept of utopia is motivated both by the material and the mystical, and indeed moves fluidly between the two? In order to understand this, it is necessary to understand something of Bloch's deeper ontology. Here I can only offer a very brief summary of key aspects of Bloch's fascinating ontology, but hopefully this will be enough to shed light on how such differently inflected conceptions of utopia can co-exist within his broader philosophy.

As mentioned in Chapter 2, the 'Not Yet' is a fundamental category for Bloch. It animates his approach to individual psychology (hence his protracted attack on Freud for failing to elaborate this aspect of the human psyche), but it also animates his metaphysics; it is central to how he views existence. In Bloch's ontology, it is not just humanity that is not yet fully actualised, but existence itself. The fundamental nature of being and reality is unfinished, the 'primordial secret' is 'still unarticulated'.[18] Even God, for Bloch, is not 'not' but rather 'not yet':

> [...] even God – as the problem of the radically new, absolutely redemptive, as the phenomenal of our freedom, of our true meaning – possesses himself within us only as a shadowy occurrence, something objectively not yet occurred [...] he 'weeps' as certain rabbis said of the Messiah, at the question, what is he doing, since he cannot 'appear' and redeem us; he 'acts' in the deepest part of all of us as the 'I am that I shall be.'[19]

Bloch also describes a mystery at the heart of existence – an inconstructable question – which is explicitly conceived in resistance to a Platonic idea of existence in which truth is eternal, the answers are eternal, and human wisdom involves moving toward this knowledge. Indeed, he blames the ongoing influence of Plato's memory-laden philosophy for obscuring the importance of the Not Yet: 'Time and again, it was ultimately the ceiling of Plato's anamnesis above dialectically open Eros which kept out and, in a contemplative antiquarian fashion, closed off previous philosophy, including Hegel, from the seriousness of the Front and the Novum.'[20] The Blochian universe is arguably both more fraught, and more creative, than the Platonic universe. Outside of the Blochian cave there is only a blur of possibility: we might intuit in the Not-Yet-Conscious a sense of what could be, but only in the act of becoming, of making, does the Not Yet realise itself. What fires all of Bloch's work is the possibility of this creative forward-dawning, whether it be in the child's hopes and dreams, societies' new formations, or metaphysical emergence.

Bloch's notion of an inconstructable question at the heart of existence can alternatively be understood as a kind of lack: the question has yet to be answered (or even conceived), the lack has yet to be filled. In a recorded conversation with Adorno, Bloch draws attention to the fundamental importance of Brecht's short sentence in *Mahagonny*: 'Something's Missing.'[21] This lack is several times characterised by Bloch as an alienation of subject from object, this 'curious gap that always lies between subject and object', and of the subject from itself.[22] And here we arrive at a further two descriptions of the utopian, outlined below.

Third, the utopian is the union of subject and object, it is the overcoming of the foundational alienation to which the utopian impulse always tends. Bloch refers to the place and time where and when there will finally be no more alienation, either material or spiritual, as home, *Heimat,* homeland. Both the nostalgic and Bloch's utopian long for 'home', then, although Bloch's home is not a place already known or remembered (however mythically). Rather, it is a place at which we have yet to arrive. Richard Gunn, in his review of Bloch's *Principle of Hope,* describes this aspect of Bloch's thought well:

> Bloch, according to whom theorising means 'venturing beyond', wagers the success of his own venturing upon a future in which every aspect of alienation – the alienation of subject from object, of self from other and of humankind from nature – is at last overcome.[23]

Fourth, and this is the conception of utopia that is closest to the core of Bloch's philosophy, the utopian is Not Yet, and the utopian impulse is that which animates the space between the 'already is' and the 'Not Yet'. The always-already-known, the dead letter, is never utopian for Bloch; the utopian impulse can never be a reaffirmation of something that already is, or once was. As Craig Hammond points out in his PhD thesis on Bloch and contemporary cinema, utopia is 'assigned the formula of the *Not-Yet,* and so, can be understood as consisting of undisclosed hints towards the new possibilities on the horizon of tomorrow'.[24] Caroline Edwards also emphasises this point: 'Bloch insists that the "Not Yet" reveals how emancipatory futural possibilities are germinative within the present through a utopian hermeneutics of longing, expectation and hope.'[25]

How much of Bloch's ontology does one have to adhere to in order to pursue a Blochian method of cultural or literary criticism? On one level: not much. It is perfectly coherent to practise a cultural criticism that, inspired by Bloch and in resistance to more common Marxist or neo-Marxist practices, does not seek merely to expose ideology but instead combs through mass cultural texts and seemingly ideological cultural practices for hopeful elements that have been warped in an alienated culture that cannot express hope directly. One need not touch on ontology, either, to see as hopeful those elements which contain resistances to capitalism, barely conscious hints of alternatives to 'what is', or a longing toward a less alienated world. However, at least some understanding of Bloch's wider philosophy is helpful in order to grasp the felt urgency, and intensity, of his formulation of the 'Not Yet'. The utopian is, for Bloch, an expression of a dawning futurity: the thrilling, Not-Yet-Conscious tentative blossoming of a sense of possibility, the possibility that things might be other than they are – less alienated, more connected, more true, more beautiful, more whole.

In Part II between every chapter in which I give a reading of a different contemporary nostalgia text, I will include a 'theoretical interlude' in which I consider further perspectives on the utopian in Bloch and his contemporaries. I will do this in order to deepen and enrich the core ideas that I have laid out here, and to give a fuller sense of what I mean in using the term utopian in connection with contemporary nostalgia culture. Now, though, I turn to Bloch's discussion of non-contemporaneity, to look at the way he applies his method of ideology critique specifically to non-contemporaneous material.

Bloch's non-contemporaneous: traces of hope

Bloch introduced the concept of *Ungleichzeitigkeit*, or non-contemporaneity (sometimes translated as non-simultaneity), in *The Heritage of Our Times*, a series of essays written throughout the 1920s and 1930s (and published in 1935) in which he analysed the emergence of fascism via readings of cultural objects and events in the Weimar republic's 'Golden Twenties'. The opening lines of the central section of *The Heritage of Our Times* (in which Bloch presents his most extensive philosophical analysis of non-contemporaneity) give a clear sense of what he means by non-contemporaneous.

> Not all people exist in the same Now. They do so only externally, through the fact that they can be seen today. But they are thereby not yet living at the same time with the others.[26]

He means, first, that at any moment in time, sub-cultures or cultural practices can and do exist whose customs, approaches, worldview, sense of self or sense of nature do not accord with the dominant culture, and belong rather, or accord with, cultures of an earlier time. This idea is similar to the notion of 'residual culture' which Raymond Williams describes as:

> reaching back to those meanings and values which were created in real societies in the past and which still seem to have some significance because they represent areas of human experience, aspiration and achievement which the dominant culture undervalues or cannot recognise.[27]

In addition to non-contemporaneous sub-cultures, Bloch also believes that one can find, within a single culture or socio-economic group, conflicting cultural currents, norms or desires, some of which are highly contemporary and reflect the dominant culture, and others which are closer in kind to those of an earlier era.

Theorist David Gross builds on Bloch's concept by distinguishing between three types of non-contemporaneity: absolute non-contemporaneity (a lost past, accessed only abstractly by historical or archaeological reconstruction), relative non-contemporaneity (lingering residues of old-fashioned codes of honour, antiquated spiritual outlooks, or scraps of ancient cosmologies) and enduring non-contemporaneity (modes of thinking and behaving that have survived intact to the present age, although where one encounters them – usually in rural enclaves, ethnic subcultures or undergrounds – they seem 'out of sync' with the time).[28]

More important, in terms of its relevance for evaluating the possible utopian content of nostalgia, is Bloch's distinction between the *subjectively* non-contemporaneous and the *objectively* non-contemporaneous.[29] By the objectively non-contemporaneous, Bloch is referring to the 'continuing influence of older circumstances and forms of production [...] as well as of older superstructures'.[30] Economic, political, and cultural reality reflects different stages of development, and declining remnants of the objectively non-contemporaneous continue to assert a real influence on certain areas and in certain ways. But by the subjectively non-contemporaneous, Bloch implies not just actual social realities, but cultural longings and desires that are out-of-kilter with the Now: a 'muffled non-desire for the Now'.[31] These subjectively non-contemporaneous longings often present themselves as cultural nostalgia, something that becomes clear if we look closer at the specific context within which and about which Bloch was writing.

Bloch's concept of non-contemporaneity is introduced via an analysis of the *Mittelstand* (the professional middle classes, civil servants, small- and medium-scale farmers and shopkeepers), or as Bloch variously calls them, the 'little people' or 'average man', in Weimar Germany.[32] He argues that these middle classes felt desires and longings that related to times and cultures gone by but persisted into the present. He talks about a 'secret Germany', contained within and alongside modern Germany, which is a 'gigantic, seething container of the past', in which mythologies of the land and a longing for a simpler life, 'a piece of mythical *enchantment by the soil*', abound.[33] He thought that these non-contemporaneous longings were still strong in parts of Germany and its middle class, despite the fact that it very much belonged to the industrial world. His descriptions of these non-contemporaneous currents evoke a sense of their power:

> rootedness has been posing here for a long time as the strangest in-rootedness, limiting fate as chthonic fate, just as if the soil itself were still saturated with ancient earth cults and held – with a kind of objective romanticism – its inhabitants tight. From this perspective, Alemannia-Bavaria seems like a Catalunian battlefield in which slaughtered myths or myth-contents are still circulating after the real battle, and not only aesthetically, but in the spell of the ancient location.[34]

To identify temporally conflicted identities, longings and cultural customs is not original, although Bloch's specific descriptions of Weimar Germany are richly evocative. What is original is Bloch's normative account of how one should understand these non-contemporaneous elements, and of what function they should and could have played. He is highly critical of the Left – of the German Communist Party (KPD) and 'vulgar' Marxism more generally – who dismissed as irrational and ideological these non-contemporaneous, nostalgic longings, on which the Nazis were thus able to capitalise. The Nazis converted these out-of-kilter romantic yearnings to further their own capitalist agenda, and divert attention from contemporary proletarian oppression, a point which is well summarised by Kaufman:

> The Nazis were brilliant in seeing how the radical anti-industrial, reactionary but still proto-revolutionary impulses of various disenchanted groups whose formations preceded and survived the growth of capitalism could be mined to maintain capitalistic growth. In other words, fascism used the energy of non-synchronous elements of German society – the hold-out from previous eras – to save the industrial state from its own contradictions and crush the leftist alternative.[35]

What makes the Nazi success in doing this particularly troubling for Bloch is that the non-contemporaneous romantic longing of the time and contemporary capitalist reality were not in fact obvious or easy bedfellows:

> But even now capitalism cannot totally keep peace with two 'truths': with an irrational one on the one hand, for its populace, with an abstract-mechanical one on the other, in the factory. [...] The blood myth, and intoxication as a whole, is not the most desirable servant of capitalist *reason*.[36]

Bloch presents the Nazi's successful exploitation of this form of cultural nostalgia as a testament both to Nazism's ingenuity and to capital's tenacity, given that it encompassed this obviously unlikely pairing. Likewise, he saw it as a clear failure on the part of the Left that these alternative currents were simply dismissed as reactionary claptrap, since this allowed the Nazis to benefit from what could have, and arguably should have, had more in common with the Left.

Bloch then applies his general method of ideology critique to the non-contemporaneous longings of Weimar Germany: he takes what the left dismissed (or exposed) as pure ideology, and mines it for the authentic utopian core, shards of latent hope. He first agrees with the conventional Marxist perspective that there was much in these longings which was simply small-minded and regressive, not to mention dangerous – 'an illogical space can develop, in an increasingly undisturbed way, in which wishes and romanticisms, primeval drives and mythicisms revive'.[37] For example,

> the insecurity which produces merely nostalgia for what has been [...] places figures in the midst of the city which have not been seen for centuries [...]. Older sorts of being thus recur [...] and older images of hate as well, like that of Jewish usury as exploitation per se.[38]

This is very important, as he is not proposing an uncritical celebration of the non-contemporaneous (as we have seen, it would be a foundational mistake for Bloch to locate utopia in the past) – and likewise in following a Bloch-inspired approach towards contemporary culture I am not proposing an uncritical celebration of nostalgia culture.

However, and this is central, Bloch *also* goes on to identify, to sift out, other elements in these longings, 'fields of a different irratio'.[39] The aspects that he picks out, in keeping with my earlier characterisation of Bloch's utopia, are those, first, which were at odds with modern capitalist society and, second, that longed towards a less alienated world. He discovers in these nostalgic longings (as well as insecurity and hatred) a kind of resistance to rationalisation, to the cold rule of technology and to the *Phantasiemord* (murder of the imagination) that pervaded modern human relations.[40] In particular, from within the non-contemporaneous, he picks out as utopian the longing for human relations that were closer and less alienated than they had become under capital's advance:

> [...] the relatively more lively aspect and wholeness of earlier relations between human beings [...]. These relations were still relatively more direct than the capitalist ones, they brought more 'matter' with them than today both in the human beings between whom they prevailed and in the environment which they acted upon.[41]

It should be noted that these aspects that Bloch picks out are merely longings; they are not actually being realised in the world. As mentioned briefly in Chapter 2, Bloch distinguishes between 'abstract' and 'concrete' utopia, which he sometimes calls a distinction between 'fraudulent hope' and 'genuine hope', between 'wishing' and 'wanting', and between 'wishful dreams' and 'waking dreams'.[42] The idea is that a mere wish, or longing, for something is not automatically utopian; it must also contain the desire and capacity to bring about real change in the world.

At a first glance, it might seem that the illogical out-of-kilter romantic-nostalgic longings in Weimar Germany were nothing other than mere wishing, that they contained no real transformative potential. However, in practice, Bloch's ideology critique is often more favourably inclined toward even seemingly wishful thinking than a sharp distinction between abstract and concrete utopia would suggest. As Levitas points out, this is in contrast to Mannheim, who 'consigns many forms of wishful thinking to the scrap-heap of ideology without a second thought'.[43] Bloch also consistently rallies against bourgeois realism, as for example when he hits back against Heidegger's dismissal of wishful thinking, arguing that 'it sounds like a eunuch accusing the infant Hercules of impotence'.[44] In his analysis of the non-contemporaneous in Weimar Germany, then, although the desires that he carefully picks out are perhaps not yet fully utopian, they can be seen as having utopian potential: they are sufficiently strongly felt and resistant to 'what is' to make them powerful, and give them a kind of latent transformative potential. If they had been appealed to more effectively, Bloch implies, history might have gone a different way.

Specifically, Bloch thought the Left had failed in not looking for this kind of hopeful content within the romantic backward-looking longings of the *Mittelstand*; and in not seeking to appeal to those latent aspects of their desires (resistance to rationalisation, a desire for closer relations, and so on). He thought that this failure was part of a wider failure of a Marxism that focused only on the 'cold stream' (cool, objective analysis), neglecting the 'warm stream' which had animated it (the liberating, humane desire to overcome alienation.)[45] This neglect of the 'warm stream' was detrimental, as Bloch argues in the following passage:

> This is a strange, a disastrous circle: precisely the capitalist factory dams up 'soul', and it seeks to flow away, indeed to explode against dreariness and dehumanization; but precisely vulgar Marxism, which employees first encounter and which is actually not that rare, cordons off their 'soul' for them again, even theoretically, and thus drives them back to a reactionary 'idealism'.[46]

Bloch, Marcuse and Derrida: ghosts which need redeeming

An important question remains: If Bloch's critical method of identifying the 'gold' within the 'gold-bearing rubble' of ideology rests on an understanding of the 'gold' (the utopian) as that which is Not Yet, how can this apply when one is searching for 'gold' (the utopian) in non-contemporaneous material? How can that which has already been, that which belongs to an earlier era, ever be 'Not Yet'? In other words, in a utopian philosophy that looks for germinative futural possibilities of emancipation, how can any hope be found in leftover remnants of the past?

In order to answer these questions, I will now take a slight detour from Bloch to reflect on Marcuse's father-son complex. In Chapter 2, I focused on the differences between Bloch and Marcuse. I drew attention to Marcuse's emphasis on an archaic memory of gratification (Eros) as informing the utopian impulse, and highlighted how this differs from Bloch, who was insistent on purging all theoretical anamnesis. However, there is a second meaning to Marcuse's *return of the repressed*, which is less commonly understood: the memory of liberation. This aspect of Marcuse's thought is much closer to Bloch's utopianism than the emphasis on the repressed memory of gratification, and can shed light on the question of how that which has already passed can ever be 'Not Yet'.

In an extended passage in *Eros and Civilization* Marcuse reflects on Freud's description of the 'primal father' in *Moses and Monotheism*.[47] Marcuse recounts Freud's description of how, in the primal horde (a temporarily functioning group prior to civilised society), the 'primal father' established domination and created an 'order' in accordance with his own interests. The hatred of the 'primal father' by his sons (whose interests were suppressed) culminates in their rebellion against the father, whom they kill. Subsequently, however, this 'brother clan' go on to deify the assassinated father and re-impose repression in the common interest of the group. And so a new order, one founded on social morality rather than overt patriarchal oppression, is born. According to Freud, an important psychological development among the 'brother clan' is the development of *guilt feeling*. He attributes this feeling of guilt to their having committed the supreme crime, patricide. The patriarch, father and tyrant, evoked feelings both of hatred and love, and so the guilt Freud describes is guilt at having murdered the loved father. It is also guilt at having unsettled the life of the group and threatened its stability (albeit a stability ensured by terror).

It is what Marcuse adds to this analysis which is most interesting. He claims that Freud's description of this sense of guilt only accounts for one of its aspects, but there was another important aspect of this guilt that Freud overlooked:

> Their revolt has, for a short span of time, broken the chain of domination; then the new freedom is again suppressed – this time by their own authority and action. Must not their sense of guilt include guilt about the betrayal and denial of their deed? Are they not guilty of restoring the repressive father, guilty of self-imposed perpetuation of domination?[48]

In other words, the initial act of patricide promised the possibility of liberation, and it is their failure to transform this into true liberation – instead reinstating at a social level the domination of the father – which is itself a crime that arouses guilt. This notion of a lost moment of possible liberation parallels the actual experience of many revolutionaries, who betray their initial uprising and the moment of possible freedom by reinstating new tyrannies. In a controversial analysis of the Christ-figure, Marcuse also reads the message of Christ as a triumph over the Father – the overthrow of Law (domination) by Agape (Eros) – but argues that the subsequent deification of the Son besides his Father was a betrayal of his own message, and of the possible moment of real freedom, by his disciples.[49] This echoes not only Bloch's views of Christ as a liberatory figure but also, more recently, Terry Eagleton and Peter Thompson's readings of a radical Christianity.[50]

Marcuse's insistence that 'civilization has to defend itself against the spectre of a world which could be free' relies, then, on a memory of liberation which, although historically betrayed, continues to make its demands on us.[51] With this in mind, one can see how nostalgia which appears at first glance to be for the past as it was, might actually be for the past as it nearly was, or could have been – that is, for lost promises and possibilities which might yet be re-awakened. In other words, although something is of or from the past, it may still contain a kernel of authentic unfinished dreams which have yet to be realised. Similarly, Bloch's approach to the non-contemporaneous requires 'an alliance which liberates from the *past* the still *possible future* within it by placing both in the present.'[52] As Wayne Hudson has observed, this possibility is actually built into the concept of *Noch Nicht* itself, since it is possible to translate *Noch Nicht* both as 'not yet' or as 'still not', indicating either something that has never been before or something that has been attempted before but is 'still not' fully realised.[53]

The 'spectre' of liberation that Marcuse mentions also has interesting parallels with the 'spectre' mentioned by Derrida in *Spectres of Marx* (1993), in which he introduces his theory of hauntology. The specific context in which Derrida was writing was Europe a few years after the fall of the Berlin Wall, Francis Fukuyama's recent proclamations of the 'end of history', and the 'end of Marxism' thereby implied. Derrida evokes Marx's own description of communism in *The Communist Manifesto* as a spectre that is haunting Europe; but rather than seeing communism as a spectre that is yet to come, as Marx had, Derrida sees it as a spectre that is not yet dead. His interest is in treating Marxism not merely as the dead letter that it was being all too readily decried, but as a ghost which can always be conjured forth, re-awakened, which can still unsettle: 'At bottom, the specter is the future, it is always to come, it presents itself only as that which could or should come back.'[54] Derrida talks specifically about what he sees happening to the study of Marx in the universities, that Marx's work is treated as a dead text, one to be studied purely for historical interest: 'We'll treat him calmly, objectively, without bias: according to the academic rules, in the University, in the library, in colloquia!'[55] This accords with David Harvey's (2000) comparison of the experience of having taught Marx in the early 1970s, when 'participation was still understood as a radical act', to his experience of teaching in the late 1990s:

> I teach *Capital* purely as a respectable regular course. [...] Most of the graduate survey courses in other departments now allot Marx a week or two, sandwiched in between, say, Darwin and Weber. Marx gets attention. But in academia, this is devoted either to putting him in his place as, say, a 'minor post-Ricardian' or passing him by as an out-moded 'structuralist' or 'modernist'.[56]

This image of the 'specter', evoked by both Marcuse and Derrida, is used also by Žižek to describe Bloch's approach to the non-contemporaneous:

> [...] these betrayed radical-emancipatory potentials continue to 'insist' as a kind of historical specter and to haunt to revolutionary memory, demanding their enactment, so that the later proletarian revolution should also redeem (put to rest) all these past ghosts.[57]

The ghostly is a helpful trope for understanding what Bloch looks for in non-contemporaneous cultural currents, because it evokes a sense that sometimes things that stick around have unfinished business, need to right remaining wrongs, are unsettled by the current state of affairs (and indeed can unsettle it). As such, rather than simply denouncing nostalgia as regressive, or stuck in the past, one might find elements that are not, at heart, about a desire to go back in time, but rather a desire to re-awaken in the present context, towards future emancipation, the betrayed promises of yesterday that still have relevance today. As Bloch puts it:

> [...] by false consciousness and unfounded romanticism being everywhere expelled, of course, but also by [...] taking in the subversive and utopian elements, the repressed matter of this not yet Past.[58]

A final point of interest is that just as Marcuse is concerned with re-imagining the father-son complex, so Derrida invokes Hamlet and the ghost of the father to articulate his concept of hauntology. A more literal point of reference is Hannah Arendt's characterisation of 'an entire generation of Jewish-German intellectuals' (Bloch's generation) as raging in a struggle between fathers and their sons.[59] The most overt battle Arendt is referring to was about the content of Judaism and the future of the Jewish people, but perhaps it also reflects Bloch and his contemporaries' inner conflict regarding how the past might inform the future more broadly.[60] The symbolic value of the father-son dynamic lies in the fact that it encapsulates the complex relationship between tradition and the novel, and, crucially, between past struggles and future hopes.

Benjamin's shock of recognition

As mentioned in the Introduction, Bloch and Benjamin were friends, and frequently corresponded about their ideas. An important similarity between them, with regard to this work, is that like Bloch, Benjamin sought to redeem the ruins of the past, to liberate from within them what may yet have utopian potential. Gilloch describes Benjamin's vast *Arcades Project*, which preoccupied him throughout the final decade of his life, as aiming 'to liberate and redeem genuine utopian moments buried in mythic forms'.[61] This fondness for excavation of the outmoded is wonderfully described by Hannah Arendt in the introduction to *Illuminations*, in a passage which is worth quoting in full:

> Like a pearl diver who descends to the bottom of the sea, not to excavate the bottom and bring it to light but to pry loose the rich and the strange, the pearls and the corals in the depths, and to carry them to the surface, this thinking delves into the depths of the past – but not in order to resuscitate it the way it was and to contribute to the renewal of extinct ages. What guides this thinking is the conviction that although the living is subject to the ruin of time, the process of decay is at the same time a process of chrystallization, that in the depth of the sea, into which sinks

> and is dissolved what once was alive, some things 'suffers a sea-change' and survive in new chrystallized forms and shapes that remain immune to the elements, as though they waited only for the pearl diver who one day will come down to them and bring them up into the world of the living – as 'thought fragments,' as something 'rich and strange,' and perhaps even as everlasting *Urphänomene*.[62]

The parallels in basic approach with Bloch are clear: Bloch's 'gold' in the 'rubble' is Benjamin's 'pearl' in the sea bed. However, I want to draw attention specifically to the role of shock in relation to forgotten utopian promise, an idea which Benjamin dwells on, and which provides an important additional angle from which to consider the theoretical perspective that this chapter has been developing.

A useful starting point is a conversation that Bloch had with Benjamin while visiting Capri in 1924 and which he recounted much later, in the 1960s. The two had apparently been discussing the phenomenon of *déjà vu*, and had concluded that rather than providing evidence of Platonic anamnesis, as is often assumed, *déjà vu* expresses an 'act of intention'.[63] Kaufman summarises the importance of this insight well:

> What has been 'seen' before is not the point: one recovers an interrupted intention, an orientation to and in the world. The eerie sensation of shock – the sense that one has been here before – is quite real and quite accurate, but it registers an orientation that has been forgotten or repressed. [...] Shock marks the recognition of what has not been fulfilled, of an intention and self-orientation that have been actively displaced and disrupted. The scary revenant in the uncanny present is not the infinite repetition of the past – it is rather a reminder of what one had meant to do for the future.[64]

What exactly is meant by this 'eerie sensation of shock' can be elaborated by looking at two of Benjamin's famous essays: 'The Image of Proust' and 'On Some Motifs in Baudelaire'.

In 'The Image of Proust' Benjamin asks whether the Proustian *mémoire involontaire* is not 'much closer to forgetting than what is usually called memory?'[65] By this he means that our sharply drawn lines of voluntary memory allow us access to only the tiniest part of all that has been experienced. What Benjamin valued most in *mémoire involontaire* was that it could awaken us from our official memory, to which we have become accustomed, and by which we are therefore not stirred. Because of the total unexpectedness of these forgotten, involuntary memories, we do not easily fit them into our routine narratives or explanations and we are thus momentarily reminded of a fuller experience; we are able to take in all the complexity and wonder of the remembered moment, rather than the numbing instrumentalised version of experience and memory that we more often inhabit.

Contrary, then, to the common reading of this most famous of nostalgics as dwelling in a past that obsesses him, Benjamin sees Proust's aim as a future-oriented awakening. He offers not reflection, but disruption, actualisation, and promise:

> *À la recherché du temps perdu* is the constant attempt to charge an entire lifetime with the utmost awareness. Proust's method is actualization, not reflection.[66]

A distinction that Benjamin called on again and again throughout his life is the distinction between two types of experience, *Erlebnis* and *Erfahrung*. He first mentions this distinction in an essay called 'Experience', written when he was twenty-one years old. He refers to *Erlebnis* as 'the mask of the adult [...] expressionless, impenetrable and ever the same. The adult has always already experienced everything: youth, ideals, hopes, woman. It was all illusion'.[67] The adults tell the young that

> youth is but a brief night (fill it with rapture!); it will be followed by grand 'experience', the years of compromise, impoverishment of ideas, and lack of energy. Such is life. That is what adults tell us, and that is what they have experienced.[68]

By contrast, he posits an alternative kind of experience, *Erfahrung*, which is full, rather than devoid, of meaning and spirit, which the young still have access to:

> We, however, know something different, which experience can neither give to us nor take away: that truth exists, even if all previous thought has been an error. Or: that fidelity shall be maintained, even if no one has done so yet.[69]

One way of understanding Benjamin's view of *mémoire involontaire* is that it can provide us with a glimpse of a remembered *Erfahrung*, unhabituated experience, as opposed to the habituated experience, *Erlebnis*, which dominates our day-to-day existence.

In a highly original move, Benjamin also transposes *mémoire involontaire* from the realm of the private interiors and individual recollections in which it dwells in Proust, to the modern cityscape. Benjamin paints a picture of modernity, and of urban life in particular, as conflicted. Like the bourgeois interiors of well-trodden drawing rooms, the city itself, Benjamin argued, was a dreamscape in which a kind of going through the motions was most often substituted for real, alert living. In 'On Some Motifs in Baudelaire' Benjamin argues that this experience of the city is typified by 'the haughty, insular bourgeois subject, who, maintaining distance and shunning contact, hurries joylessly past to seek refuge in exclusive cultural spaces or private interiors'.[70]

Even the newspapers read by such a bourgeois subject, Benjamin argues, act as a barrier to actual engagement with the world:

> If it were the intention of the press to have the reader assimilate the information it supplies as part of his own experience, it would not achieve its purpose. But its intention is just the opposite, and it is achieved: to isolate what happens from the realm in which it could affect the experience of the reader.[71]

Thus, on the one hand, as Gilloch puts it, 'the city is home to an amnesia born of sensory over-stimulation and fatigue, a forgetfulness which leads to a misrecognition of the always-the-same as the ever-new'.[72] That is to say that most of the time we do not experience things fully; instead we manage the intensity of modern existence by clinging desperately to the narrative of the always-already-known (we are not, in other words, open to what Bloch calls the 'inconstructable question' of existence).

On the other hand, Benjamin also sees another possible side to modern metropolitan life, which is that it can expose the members of the amorphous crowd to the shock of the unknown, thereby momentarily awakening them. This aspect of urban existence is articulated profoundly, Benjamin thought, in Baudelaire's famous sonnet, 'À Une Passante', or 'To a Passer-By'. The poem describes an arresting moment in which the poet-persona passes a woman on the street and is struck by feeling for her ('O you I would have loved, o you who knew it too!').

In this passing, erotic encounter, Benjamin sees a 'moment of enchantment' provided by the 'figure of shock', that cuts through the otherwise numbing experience of the metropolis, and provides a fleeting 'fragmentation of a coherent experience (*Erfahrung*)'.[73] Svetlana Boym suggests in her analysis that 'the poem is a nostalgia for what could have been' and that 'in one moment he (the poet) experiences a new birth and death, a pleasure and pain, darkness and light, the present and the eternal'.[74] There is also in the poem a sense of recognition: of seeing something in one another, or perhaps just of seeing one another, in a way that the usual joyless hurrying of the bourgeois subject does not allow.

The experience of the poet-persona, which Benjamin describes as 'love – not at first sight, but at last sight', rests on its transience, its impossibility.[75] Boym describes the unknown woman as 'an allegory of *modernité*', but it is rather the experience of the encounter which is an allegory of modernity, in which so much of what is available to us to experience passes us by as ephemera, and in which we are somehow unable – except in arresting moments which purport, but ultimately fail, to stop time – to genuinely experience most of what confronts us. 'À Une Passante' thus reflects the possibility for urban existence to offer moments of shock when, just as in Proust's *mémoire involontaire*, our habituated and carefully systematised consciousness is rocked by a glimpse of a different kind of experience (*Erfahrung*).

In addition to such shock encounters in a crowd, Benjamin saw another source of potential awakening in the forgotten past of the city. *The Paris Arcades* is an attempt to excavate lost utopian traces from once-fashionable Paris. He saw commodities as 'thwarted wish images' and looked for the 'unfulfilled utopian impulses' and promises embedded in them.[76] Crucially, what makes this possible for Benjamin, what gives the outmoded its power, is precisely its irrelevance. There is a genuine utopian core to present-day wishes and cultures as well, but they are too often co-opted into a utilitarian logic, seen only for their use-value, and any other aspects are thus discounted or hidden from view. Once this blinding spell has been broken, once something is no longer useful, or fashionable, it is easier, Benjamin believes, to see it from outside the dreamscape of false consciousness, to momentarily recognise in it something of its original promise. Another way to put this is to say that once something is irrelevant it is easier to see how original promises and hopes were 'thwarted' by consumer logic.

This notion of recognition brings to mind a series of taped conversations between Bloch and Michael Landmann in 1968, in which Bloch distinguishes *anamnesis* (recollection) from *anagnorisis* (recognition).[77] Bloch is here drawing on the ancient Greek, Aristotelian, notion of anagnorisis: that, one can have moments of insight, when the truth of a moment or situation is suddenly recognised, as when Oedipus recognises suddenly that his wife is his mother, or when Electra recognises that Orestes is her brother. Geohegan draws attention to the creative aspect of this jolt of recognition:

> In anagnorisis memory traces are reactivated in the present, but there is never simple correspondence between past and present. Recognition is a creative shock, where an element from the past jolts consciousness out of joint and thereby helps in the creation of novelty.[78]

Although Bloch only briefly mentions this notion, the idea of a shock of rejuvenation or recognition is more highly developed in Benjamin. Benjamin thereby adds to Bloch's critical approach to the non-contemporaneous two things: first, his emphasis on irrelevance as liberating (when things are no longer useful we are more likely to see in them something of their original promise); and, second, the notion that one of values of the non-synchronous is its ability to shock. Benjamin reminds us that the power of present-day false consciousness and instrumental reason is such that alternative possibilities for existence (or resistances to present-day ideology) are not easily experienced; only by somehow lowering the guard of a consciousness that is trained in modern life to screen against excessive stimuli (*Erlebnis*) might we be able to grasp truth- or hope-content; that is, to experience things fully (*Erfahrung*). The non-synchronous, precisely because of its non-synchronicity, can appear uncanny, it can estrange rather than console, and thus momentarily shock us into an experience that is not pre-configured into what we think we know about the world. The shock of recognising that which has not been fulfilled, forgotten orientations and hopes, is also therefore a shock of rejuvenation: a call to wake up from the semi-conscious stupor that modern man too often inhabits.

This aspect of Benjamin's thought raises important questions to consider in relation to contemporary nostalgia culture. Amongst the ruins and faded dreams of washed-up culture, among the outdated, marginal and obscure, might one be jolted by a *sudden shock of recognition* of once-powerful hopes and promises that linger, latently, like ghosts that have not yet vanished, that might yet unsettle? Just as Benjamin was mesmerised by the outmoded Paris arcades, do today's nostalgics hear in the tinny tunes of outmoded TV programmes or see in the emptied-out fashionable spots of yesterday a truth that is obscured by relevance?

Recent complementary approaches

Amongst recent contributions to nostalgia studies, Svetlana Boym's 2001 book *The Future of Nostalgia* is one of the best known. In it, Boym proposes a distinction between two kinds of nostalgia, which she calls 'restorative nostalgia' and 'reflective nostalgia'. Restorative nostalgia is nostalgia that desires to go back in time, to rebuild the lost home exactly as it was. These nostalgics, Boym observes, do not think of themselves as nostalgic, but believe their project is about truth.[79] They are driven by anxiety about historical incongruities and fear any inconsistencies in the 'wholeness and continuity of the restored tradition'.[80] Crucially, 'restorative nostalgia has no use for the signs of historical time – patina, ruins, cracks, imperfections'.[81]

Reflective nostalgia, by contrast, 'does not pretend to rebuild the mythical place called home', instead it 'dwells in *algia*, in longing and loss, the imperfect process of remembrance.'[82] An example Boym gives of this kind of nostalgia is that exhibited in the homes of ex-Soviet immigrants in New York and Boston, whom she interviews and whose homes she photographs. She describes shelves lined with nostalgic objects, 'outdated calendars with pictures of familiar wintry landscapes' and 'old wall clocks, once elegant but no longer functional, purchased somewhere at a yard sale'.[83] She tells how each apartment 'becomes a personal memory museum' and each immigrant 'becomes an amateur artist in everyday life'.[84] They are emigrés, lost between here and there, seeking through narrative creation to make sense of fragmented stories and scraps of self. These nostalgic collections, Boym argues, seek not simply to re-create home but to tell the complex and conflicted story of exile.

Boym's approach to reflective nostalgia shares much with the approach I have elaborated in this chapter. She sees such nostalgia as both retrospective and prospective, arguing, like Bloch and Benjamin, that nostalgia is 'not always for the ancient regime or fallen empire but also for the unrealised dreams of the past and visions of the future that became obsolete [...] for unrealised possibilities, unpredictable turns and crossroads'.[85] Her idea that 'one is nostalgic not for the past the way it was, but for the past the way it could have been. It is this past perfect that one strives to realise in the future'[86] clearly parallels Bloch's emphasis on the *unfinished* remnants of the non-contemporaneous, on the unredeemed ghosts of the past, haunting us from the side-alleys of non-history.

Importantly, she sees a creative forward-dawning impulse at play in reflective nostalgia: 'Creative nostalgia reveals the fantasies of the age, and it is in those fantasies and potentialities that the future is born.'[87] The tension between loss and renewal, creatively and at times tensely navigated, also recalls Bloch's response to Lukács in the debate about Expressionism:

> one must ask: *are there not dialectical links between growth and decay?* Are confusion, immaturity and incomprehensibility always and in every case to be categorized as bourgeois decadence? Might they not equally – in contrast with this simplistic and surely unrevolutionary view – be part of the transition from the old world to the new?[88]

For both Bloch and Boym our nostalgic longings and non-contemporaneous desires tell a story both about where we have come from and where we are going. Both Bloch and Boym recognise the importance of the unrealised past, of still *latent* unfinished dreams, of the slow and careful transformation of deeply felt losses, and of the creative collision of memory and possibility.

However, there are two important pieces missing from Boym's analysis of nostalgia. First, her approach distinguishes 'restorative' from 'reflective' nostalgia on the understanding that most instances of nostalgia fall into one category or the other. Her task, then, is to identify which instances of nostalgia have critical promise and which kinds are socially useful; and also to identify grounds on which these instances or kinds of nostalgia can be defended. Others who follow a similar line to Boym include Tannock, who distinguishes between 'retreat' and 'retrieval', and Pickering and Keightley, who make a distinction between 'ontological security' and 'renewal'.[89]

Since these approaches are all, to varying degrees, reacting against the deployment of nostalgia by right-wing, conservative groups, they risk simply stating that nostalgia, when used by the right wing, is bad, but when used by the left wing can be good. Tannock, for example, defends the 'back to basics' environmental movement.[90] None of these approaches allows for the possibility that instances of nostalgia cannot all be categorised as one kind or the other – bad, restorative, nostalgia or good, reflective, nostalgia; that, in other words, one instance of nostalgia might contain within it conflicted longings and possibilities. The same sentiment or cultural product can be both regressive and hopeful, ideological at the same time as containing resistances to ideology. This makes it important to do the complicated work of drawing out the utopian potential from within seemingly regressive phenomena, and it is here that Bloch's critical method is hugely helpful.

The second problem with Boym's account is that, although written in 2001, it barely engages with Jameson's critique of the nostalgia mode. Similarly, Pickering and Keightley do mention Jameson but only fleetingly. They argue that Jameson confines nostalgia to such mass cultural tendencies as surface style, stereotype, kitsch and pastiche, as these might be found in historical tourism or advertising culture, which they dismiss as merely superficial instances of nostalgia.[91] Given the sheer extent of these mass cultural tendencies (they are also identifiable in film, music and literature) it seems insufficient to merely dismiss them out of hand as uninteresting, and to discuss contemporary nostalgia culture without engaging with Jameson's damning and in many ways convincing critique. It is one thing to challenge the Jamesonian characterisation of nostalgic culture in America since the 1970s, as my readings will do; it is another thing to ignore it all together, along with the boom in commercial and mediatised nostalgia since the 1970s, thus treating contemporary forms of nostalgia as exactly the same psychological longing that so obsessed Proust and his contemporaries. My own approach in Part II is not to ignore the very mode-ish nature of contemporary cultural nostalgia, or to try to identify certain instances of nostalgia as reflective and others as unreflective. Instead, I will look anew at precisely those mass and seemingly superficial tendencies that concern Jameson.

Another recent approach worthy of serious consideration is Helmut Illbruck's theoretical defence of nostalgia as utopian, which draws on Bloch. As already mentioned, the concept of a homeland (*Heimat*) in which all alienation is overcome is very important to Bloch, who describes this homeland as 'something which shines into the childhood of all and in which no one has yet been'.[92] It is in relation to this deep philosophical concept of *Heimat* that Illbruck constructs his defence of nostalgia.

Illbruck's basic argument is that: 'In valorizing a utopian home shining into a childhood not yet attained, Bloch's philosophy also spells out the cost involved in any incredulous or distrustful farewell to nostalgia.'[93] For Illbruck, then, Bloch's major insight is to rescue the concept of 'home' – an idea that is often under attack for its seeming totality – and deem it a necessary horizon. As Illbruck says of Bloch:

> He was a lone voice in the wilderness among Marxist thinkers with his insistence that 'home' was too important a concept simply to be abandoned as the ideological product of false consciousness.[94]

For Illbruck, the home that the nostalgic longs for is more than home in any limited, personal, or even historical or cultural sense (a lost particular): it is that final home, the 'transcendental home' to which Lukács refers (see Chapter 1), home as ultimate belonging. Bloch's rescuing of 'home' implies that nostalgia is not pointless escapism or false consciousness, that it has at its heart the desire for something that is actually of paramount and ongoing importance. Bloch, in refusing to 'let the concept of *Heimat* be appropriated solely by totalitarian political movements' also, Illbruck implies, implicitly refuses to let nostalgia (which at core is nothing other than a longing for home – in Illbruck's reading, a kind of spiritual or ultimate home) be dismissed as nothing other than a conservative, ideological sentiment.

In order to do this, Illbruck contends, Bloch re-conceives that which the nostalgic longs for as being in the future rather than in the past, thus leading to the 'potentially liberating insight' that:

> the 'conservative' yearning to uphold our belonging to a tradition in which we may feel at home must nonetheless proceed on a new self-understanding, revolutionising the sense of belonging itself. Inevitably, the anti-revolutionary impetus must nonetheless re-conceive tradition in such a way that it points to a still to be imagined future and future belonging.[95]

Illbruck's philosophical analysis is insightful and complementary to my own theoretical analysis, and the attention he gives to the significance of 'home' both in nostalgia and in Bloch's utopia is important. However, his primary motivation seems to be salvaging the philosophical notion of 'home' in a postmodern landscape. He is less concerned with the myriad of actual expressions of nostalgia, either historically or in contemporary culture, and the question of how to read these expressions. As such, his theorisation of nostalgia in relation to Bloch is for the most part differently oriented to my own.

Closer to my own efforts are those by a lively community of media and critical theorists who draw on Jacques Derrida's concept of 'hauntology' to think about contemporary cultural production and its relationship to the past. Many of these theorists are bloggers, as a 2011 *Guardian* article observes: 'Hauntology is probably the first major trend in critical theory to have flourished online.'[96]

The specific interests of these critics and bloggers are different to my own. They are primarily concerned with British, as opposed to American, culture, and they focus particularly on music, which I do not consider at all. They do, however, make similar observations regarding the prevalence of cultural nostalgia and ask similar questions about how this should be interpreted. Mark Fisher (a.k.a. k-punk, his online pseudonym) is one among several theorists who have observed an abundance of cultural nostalgia in the twenty-first century. In his 2013 book *Ghosts of My life: Writings on Depression, Hauntology and Lost Futures*, he puts his case starkly:

> While 20th century experimental culture was seized by a recombinatorial delirium, which made it feel as if newness was infinitely available, the 21st century is oppressed by a crushing sense of finitude and exhaustion. It doesn't feel like the future, Or, alternatively, it doesn't feel as if the 21st century has started yet. We remain trapped in the 20th century.[97]

He argues that contemporary artists tend to rely on old styles, and that genuine innovation is rare: 'the current moment is in the grip of a *formal* nostalgia'.[98] He also claims that there has been an accordant lessening of expectations:

> There can be few who believe that in the coming year a record as great as, say, the Stooges' *Funhouse* or Sly Stone's *There's a Riot Goin' On* will be released. Still less do we expect the kind of ruptures brought about by The Beatles or Disco.[99]

For Fisher, contemporary nostalgia is directly related to an inability to produce anything artistically, or formally, innovative – a consequence, he argues, of the absolute triumph of capitalism. Other online critic-theorists have also pointed out the prevalence of nostalgic style as a sign of something amiss. For example, James Bridle of booktwo.org remarks:

> I am so bored of nostalgia. Of letterpress and braces and elaborate facial hair. I appreciate these things, but I think there's something wrong with a culture that fetishizes them to the extent that we currently do.[100]

The use of hauntology in relation to contemporary nostalgia by several of these critics, most notably Mark Fisher, has similarities with the method I will pursue. The theoretical crossovers between Derrida's hauntology, Marcuse's spectres of liberation and Bloch's persistent traces of 'gold' have already been discussed, but the parallels in method are also evident in practice.

Fisher, for example, looks back longingly to the music, films and literature of his youth (as well as other more recent cultural works), particularly to cultural works he identifies with 'popular modernism', and tries to reignite the lost futures and failed promises that he believes they contained. He says of his own method:

> What should haunt us is not the *no longer* of actually existing social democracy but the *not yet* of the futures that popular modernism trained us to expect, but which never materialised. These spectres – the spectres of lost futures – reproach the formal nostalgia of the capitalist realist world.[101]

In a similar vein, Owen Hatherley looks to early and mid-twentieth century modernism in architecture, film and theatre, and suggests:

> What might be at work here is the common contemporary phenomenon of nostalgia for the future, a longing for the fragments of the half-hearted post-war attempt at building a new society, an attempt that lay in ruins by the time I was born.[102]

He then remarks:

> We have been cheated out of the future, yet the future's ruins lie about us, hidden or ostentatiously rotting. So what would it mean, then, to look for the future's remnants? […] Can we, should we, try and excavate utopia?[103]

The criticism that these and similar efforts have often faced is that they are themselves nostalgic. Although they despair at the prevalence of nostalgia culture today and at the disappearance of the new, or the future, from our cultural terrain, they nonetheless appear nostalgic for a time when the future was more hopeful. Nostalgic, in other words, for a time when there was less nostalgia. Mark Fisher idolises in particular the popular modernism of 1970s Britain as imperfect certainly, but still 'better than neoliberalism wants us to remember' because that 'culture was opened up to working class inventiveness in a way that is now scarcely imaginable to us'.[104] Likewise, Owen Hatherley looks at post-war British modernist architecture with an undeniable fondness, believing that it once held a revolutionary promise as an 'Eldorado for the working class' which has since vanished.[105]

Charles Beckett, of www.howtothinkaboutthefuture.com suggests that 'at its simplest, then, hauntology is another form of nostalgia, a nostalgia for ideas of the future that have been rendered obsolete by the march of time and the quote-unquote "End of History"'.[106] James Bridle of booktwo.org is also sceptical, seeing hauntology as too close to the nostalgia it purports to critique for comfort. He concludes: 'I think my problem with hauntology is that it deals with the problem of the future by going back to the past. And that is fine: but it will not save us.'[107]

This criticism is not entirely warranted: as discussed, the stated approach of these critics is not just to go back to the past but to excavate utopia from past forms, to re-engage traces of hope that are *still* powerful, an approach similar to my own. However, given the different content of their past to the pasts haunting the contemporary American, often female-led, contemporary cultures that this book explores (they, and indeed I myself, have no particular attachment to 1970s British popular culture, for example) it is easy to see how personal their predilections are, how enamoured they are of the faded culture they mine for traces of hope. Fisher's work in particular, although he dislikes and resists the charge of nostalgia ('It seems strange to have to *argue* that comparing the present unfavourably with the past is not automatically nostalgic in any culpable way'), is steeped in a deeply personal love of the culture he goes back to, as suggested by the title of his book, *Ghosts of My Life*.[108]

Certainly, one should not fall prey to assuming that a purely objective stance in criticism is ever possible; perhaps it is not even desirable. As Adorno reminds us:

> To anyone in the habit of thinking with his ears, the words 'cultural criticism' (*Kulturkritik*) must have an offensive ring, not merely because, like

> 'automobile', they are pieced together from Latin and Greek. The words recall a flagrant contradiction. The cultural critic is not happy with civilization, to which alone he owes his discontent. He speaks as if he represented either unadulterated nature or a higher historical stage. Yet he is necessarily of the same essence as that to which he fancies himself superior.[109]

Moreover, as Adorno insists, 'The complicity of cultural criticism with culture lies not in the mere mentality of the critic': it is not, in other words, something that we can easily think ourselves out of.[110] Despite these inherent difficulties, one can at least strive to avoid becoming overly enamoured of one's subject, as Fisher is, if doing so blinds you to its problems, or to the possibilities inherent in other cultures that are of less personal appeal. There is a fine line, after all, between mining nostalgia culture for traces of utopia, of lost promises that are *still* powerful, and using this approach as an excuse to dwell in and on a romanticised past. In other words, we must aim, as far as is possible, to avoid merely reflecting the nostalgia culture that we seek to better understand.

Introducing Part II

In this chapter, I have elaborated my core Bloch-inspired theoretical approach to nostalgia culture. In summation, it could be described as the search for gold in the nostalgic rubble; but this chapter has elaborated in depth the reasons for pursuing this method of cultural criticism.

In Part II I will give close readings of three contemporary nostalgia texts. In these, I demonstrate this Bloch-inspired method of cultural criticism and way of reading nostalgia, which I hope will lead to a deeper understanding of contemporary nostalgia culture than that provided by existing Jameson-inspired analyses. In each chapter, I will first show how the text in question could, at first glance, be read as exemplary of Jameson's 'nostalgia mode'. These readings, if taken alone, would dismiss the nostalgia texts in question as merely ideological reflections of depthless late capitalism. Next, I will give a reading of the text that adds nuance to that picture – one which looks for 'The Gold' in the rubble. Here I will attempt to identify latent utopian potentials (ghosts which need redeeming) in the texts in question. It is important to note that neither reading alone would be sufficient: I do not highlight the ways in which these texts are indeed depthless products of late capitalism to then entirely disprove this perspective – it undoubtedly contains some truth content. But it is not the whole picture. Similarly, the subsequent search for the 'gold' in the 'rubble' is not an attempt to de-fang Bloch (who himself was highly critical of then contemporary and capitalist cultures), or to reappropriate him as superficially hopeful. It recognises the merit of the former readings of these texts as ideological, but then attempts to *also* do the difficult work of uncovering hope-content in these cultural products – of examining the unspoken longings that animate them.

I began this chapter by putting forward four possible conceptions of the utopian which, taken together, serve as a guide for our present task: identifying aspects of each nostalgia text that have utopian potential. However, as already mentioned, defining utopia is no easy matter and these four suggested guidelines will only take us so far. Each chapter in Part II is therefore preceded by a 'theoretical interlude'. The purpose of these theoretical interludes is to expand the notion of *utopian* that we are working with, by exploring additional perspectives on utopia offered by Bloch and his contemporaries. The particular utopian aspect considered in each theoretical interlude will be especially relevant to the text under discussion in the chapter that follows. These theoretical interludes enrich the core ideas that have been elaborated already. As such, they should be read as a cumulative continuation of the development of ideas put forward in this chapter, and also as prompts for the analyses of the chapters in Part II.

Notes

1 Jameson, *Marxism and Form*, p. 128.
2 Bloch, *The Principle of Hope*, I, p. 59.
3 In Bloch's survey of social utopias (*The Principle of Hope*, II, pp. 471–625) he is highly critical of utopias located in a past 'Golden Age'. He is also, it should be noted, critical of utopias located in an impossible future or a remote wonderland, which are deemed guilty of neglecting the really possible Novum by painting utopia as something un-realisable (see Chapter 1).
4 Douglas Kellner, 'Ernst Bloch, Utopia and Ideology Critique', in *Not Yet: Reconsidering Ernst Bloch*, ed. by Jamie Owen Daniel and Tom Moylan (London: Verso, 1997), pp. 80–96 (p. 82).
5 Kellner, p. 82.
6 Raymond Williams, 'Base and Superstructure in Marxist Cultural Theory', *New Left Review*, 1.82 (1973), 3–16 (p. 12).
7 Kaufman, p. 37.
8 Kellner, p. 83.
9 Kellner, p. 85.
10 Kellner, p. 86.
11 Jameson, *Marxism and Form*, p. 121.
12 Jameson, *Marxism and Form*, p. 121.
13 Giles Lipovetsky, *Hypermodern Times*, trans. by Andrew Brown (Cambridge: Polity Press, 2005), p. 59.
14 Ernst Bloch, *Heritage of Our Times*, trans. by Neville and Stephen Plaice (Cambridge: Polity Press, 1991), p. 116. Caroline Edwards has also drawn attention to this phrase from *The Heritage of Our Times*, in her chapter, 'Uncovering the "Gold-Bearing Rubble"': Ernst Bloch's Literary Criticism'.
15 Žižek, Preface, *The Privatization of Hope*, p. xix.
16 Bloch, *Heritage of Our Times*, p. 115, p. 113 (my italics).
17 Bloch, *Heritage of Our Times*, p. 108.
18 Bloch, *The Spirit of Utopia*, p. 194.
19 Bloch, *The Spirit of Utopia*, p. 201.
20 Bloch, *The Principle of Hope*, I, p. 18.
21 Ernst Bloch, *The Utopian Function of Art and Literature: Selected Essays*, trans. by Jack Zipes and Frank Mecklenburg (Cambridge, MA: MIT Press, 1988), p. 15.

22 Bloch, *The Principle of Hope*, I, p. 296.
23 Richard Gunn, 'Ernst Bloch's *The Principle of Hope*', *Edinburgh Review*, 76 (1987), 1–9 (p. 2).
24 Hammond, p. 14.
25 Edwards, 'Uncovering the "Gold-bearing Rubble"', p. 188.
26 Bloch, *Heritage of Our Times*, p. 95.
27 Williams, p. 12.
28 Gross, pp. 142–143.
29 Bloch, *Heritage of Our Times*, pp. 108–109.
30 Bloch, *Heritage of Our Times*, p. 108.
31 Bloch, *Heritage of Our Times*, p. 108.
32 Edwards, 'Uncovering the "Gold-bearing Rubble', p. 183; Bloch, *Heritage of Our Times*, pp. 14, 16.
33 Bloch, *Heritage of Our Times*, pp. 51, 49.
34 Bloch, *Heritage of Our Times*, p. 50.
35 Kaufman, p. 35.
36 Bloch, *Heritage of Our Times*, p. 55.
37 Bloch, *Heritage of Our Times*, p. 102.
38 Bloch, *Heritage of Our Times*, p. 101.
39 Bloch, *Heritage of Our Times*, p. 103.
40 Edwards, 'Uncovering the "Gold-bearing Rubble', p. 189.
41 Bloch, *Heritage of Our Times*, p. 111.
42 Bloch, *The Principle of Hope*, I, pp. 5, 46.
43 Levitas, p. 71.
44 Bloch, *The Principle of Hope*, I, p. 145.
45 Bloch, *The Principle of Hope*, III, p. 1369.
46 Bloch, *The Principle of Hope*, I, p. 53.
47 Marcuse, *Eros and Civilization*, pp. 60–70. The following paragraph offers a brief summary of the key argument in this extended passage.
48 Marcuse, *Eros and Civilization*, p. 67.
49 Marcuse, *Eros and Civilization*, p. 70.
50 See Ernst Bloch, *Atheism in Christianity*, trans. by J. T. Swann (London: Verso, 2009); Peter Thompson, 'Introduction' in *Atheism and Christianity* (London: Verso, 2009); and Terry Eagleton, *Reason, Faith and Revolution: Reflections on the God Debate* (New Haven: Yale University Press, 2009). Eagleton, for example, remarks:

> What is at stake here is not a prudently reformist project of pouring new wine into old bottles, but an avant-gardist epiphany of the absolutely new – of a regime so revolutionary as to surpass all image and utterance, a reign of justice and fellowship which for the Gospel writers is even now striking into this bankrupt, dépassé, washed-up world.
>
> p. 23

51 Marcuse, *Eros and Civilization*, p. 93.
52 Bloch, *Heritage of Our Times*, p. 110.
53 Wayne Hudson, *The Marxist Philosophy of Ernst Bloch* (London: Macmillan, 1982), pp. 19–20.
54 Jacques Derrida, *Spectres of Marx: The State of the Debt, the Work of Mourning and the New International*, trans. by Peggy Kamuf (New York: Routledge, 2006), p. 48.
55 Derrida, *Spectres of Marx*, p. 38.
56 David Harvey, *Spaces of Hope* (Edinburgh: Edinburgh University Press, 2000). p. 3, pp. 4–5.

57 Žižek, Preface, *The Privatization of Hope*, p. xix.
58 Bloch, *Heritage of Our Times*, p. 114.
59 Hannah Arendt, 'Introduction: Walter Benjamin: 1892–1940', in *Illuminations: Essays and Reflections*, by Walter Benjamin, trans. by Harry Zohn (New York: Schocken Books, 1969), pp. 1–59 (p. 26).
60 Martin Jay, *The Dialectical Imagination: A History of The Frankfurt School and the Institute of Social Research, 1923–1950* (Berkeley: University of California Press, 1996), p. 35.
61 Gilloch, p. 127.
62 Arendt, p. 51.
63 Kaufman, p. 35.
64 Kaufman, p. 36.
65 Walter Benjamin, 'The Image of Proust', in *Illuminations: Essays and Reflections*, trans. by Harry Zohn (New York: Schocken Books, 1969), pp. 201–217 (p. 202).
66 Benjamin, 'The Image of Proust', *Illuminations*, p. 211.
67 Walter Benjamin, 'Experience', in *Selected Writings*, ed. by Marcus Bullock and Michael W. Jennings, vol. 1 (Cambridge, MA: Harvard University Press, 1996), pp. 3–6 (p. 3).
68 Benjamin, 'Experience', *Selected Writings*, p. 3.
69 Benjamin, 'Experience', *Selected Writings*, p. 4.
70 Gilloch, pp. 95–96.
71 Walter Benjamin, 'On Some Motifs in Baudelaire', in *Illuminations: Essays and Reflections*, trans. by Harry Zohn (New York: Schocken Books, 1969), pp. 155–201 (p. 158).
72 Gilloch, pp. 216–217.
73 Gilloch, p. 208.
74 Boym, *The Future of Nostalgia*, p. 21.
75 Benjamin, 'On Some Motifs in Baudelaire', *Illuminations*, p. 169.
76 Gilloch, p. 127.
77 Michael Landmann, 'Talking with Ernst Bloch: Korčula, 1968', *Telos*, 25 (1975), 165–185.
78 Geohegan, p. 37.
79 Boym, *The Future of Nostalgia*, p. 41.
80 Boym, *The Future of Nostalgia*, p. 45.
81 Boym, *The Future of Nostalgia*, p. 45.
82 Boym, *The Future of Nostalgia*, pp. 50, 41.
83 Boym, *The Future of Nostalgia*, p. 327.
84 Boym, *The Future of Nostalgia*, p. 327.
85 Boym, *The Future of Nostalgia*, p. xvi.
86 Boym, *The Future of Nostalgia*, p. 351.
87 Boym, *The Future of Nostalgia*, p. 351. Boym's casual identification in this one passage of reflective nostalgia with mourning is helpful insofar as it draws attention to the creative tension between the past and the future evident in both. However, it is a comparison which should not be taken too far. There are important differences between nostalgia, mourning and indeed melancholia, and although the relationship between the three is intriguing, to do it any justice would require a detailed consideration in light of a range of psychoanalytic theories.
88 Ernst Bloch, 'Discussing Expressionism', in *Aesthetics and Politics*, ed. by Ernst Bloch *et al.* (London: Verso, 2007), pp. 16–28 (p. 23).
89 Tannock, p. 458; Pickering and Keightley, p. 921.
90 Tannock, p. 455.

91 Pickering and Keightley, p. 924.
92 Bloch, *The Principle of Hope*, III, p. 1376.
93 Helmut Illbruck, *Nostalgia: Origins and Ends of an Unenlightened Disease* (Evanston, IL: Northwestern University Press, 2012), p. 233.
94 Illbruck, p. 234.
95 Illbruck. p. 235.
96 Andrew Gallix, 'Hauntology: A Not-So-New Critical Manifestation', *Guardian*, 17 June 2011, www.theguardian.com/books/booksblog/2011/jun/17/hauntology-critical [accessed 13 May 2015].
97 Fisher, *Ghosts of My Life*, p. 8.
98 Fisher, *Ghosts of My Life*, p. 8.
99 Fisher, *Ghosts of My Life*, p. 8.
100 James Bridle, 'Hauntological Futures', http://booktwo.org/notebook/hauntological-futures/ [accessed 10 January 2016].
101 Fisher, *Ghosts of My Life*, p. 27.
102 Owen Hatherley, *Militant Modernism* (Winchester: O-Books, 2008), p. 8.
103 Hatherley, *Militant Modernism*, p. 3.
104 Fisher, *Ghosts of My Life*, p. 25, p. 26.
105 Hatherley, *Militant Modernism*, p. 6.
106 Charles Beckett, 'How To Think About The Future', www.howtothinkaboutthefuture.com/?p=75 [accessed 10 January 2016].
107 Bridle, 'Hauntological Futures'.
108 Fisher, *Ghosts of My Life*, p. 25.
109 Theodor W. Adorno, 'Cultural Criticism and Society' in *Prisms*, by Theodor W. Adorno, trans. by Samuel and Shierry Weber (Cambridge, MA: MIT Press, 1983), pp. 17–35 (p. 19).
110 Adorno, 'Cultural Criticism and Society', p. 22.

87 [illegible]
88 [illegible], p. 178.
89 [illegible] p. 113.
90 Ibid.

Part II

Interlude 1

The hole in the whole: utopia contra instrumental reason

In an essay written in his early twenties, 'The Lives of Students' (1915), Benjamin discusses how most university 'education' simply prepares people for an unquestioning, fated life, rather than opening them up to real and challenging thought. The essay is striking to read a century on, as that which Benjamin bemoaned in his own day is even more characteristic of universities now; but it is also an excellent example of the deep disdain for instrumental reason that both Benjamin and Bloch shared (along with most of their Frankfurt School contemporaries). Benjamin comments:

> The concept of 'science' or scholarly disciplines (Wissenschaft) serves primarily to conceal a deep-rooted, bourgeois indifference [...] for the vast majority of students, academic study is nothing more than vocational training.[1]

At a later point in the essay, he adds:

> All these institutions are nothing but a marketplace for the preliminary and provisional, like the bustling activity in lecture halls and cafes; they are simply there to fill the empty waiting time, diversions from the voice that summons them to build their lives with a unified spirit of creative action, eros and youth.[2]

The parts of the essay in which Benjamin lampoons the so-called liberal students whose 'views have not developed beyond the level of the liberal press' reads as a scathing critique of left(ish) culture today; and his descriptions of the fun-insistent student culture – their 'few years of bourgeois freedom', 'a dearly purchased state of intoxication [...] arises between a squandered youth and a bought-out old age' – would be apt to describe any one of the thousands of student bars that surround American campuses today.[3]

The underlying point, however, is that Benjamin's essay makes apparent the depth of contempt that he felt for instrumental reason and the devastating hold it had on modern life, even in institutions of supposed learning. This is a contempt shared equally by Bloch, who argued that directing attention towards what is 'known', what is immediately explicable within a framework of causality, is to do a fundamental disservice to the possibility of actual learning, which is necessarily motivated by amazement:

> [...] man, in other words, if he finds himself just once on the stage of this extrasubjectively illuminated, extrasubjectively ascending science or pyramind of ideas, *forgets the question originally motivated by self-amazement*, indeed unthinkingly allows the Encyclopedia, the plaintiff having disappeared, to present itself as the day of absolute maturity and the comprehensive answer. Just so does that senseless game persist of 'the universal enigma' and especially its 'answer', which honest philosophers are supposed to only suggest, but which every philosopher till now, *reducing the primordial question to a mere scholastic problem*, the ad hoc problem of a system's beginning, affects to have located in some monumental term. But one should understand here and remember: by its very nature, the solution, at least, will never be expressed in a book, and as little in a mundane church as in any academic philosophy.[4]

For both Bloch and Benjamin, an important aspect of the utopian is that it resists instrumental reason. Benjamin, for example, clings to the possibility that, despite the increasing success of the university 'in ridiculing the few surviving visionaries as starry-eyed dreamers', the dreamers will continue to strive for 'that expansive friendship between creative minds, with its sense of infinity and concern for humanity as a whole'.[5] The hope is that they 'will succeed in liberating the future from its deformed existence in the womb of the present.'[6] For Bloch, too, the utopian defies the ways in which:

> human beings collapse into themselves, without a path or a goal beyond the quotidien. They lose their properly human wakefulness, substantiveness, existence [...] and finally everything grand, powerfully massive, atomizes under the 'knowing' gaze into false, disenchanted details; every blossoming becomes a whitewash, or ultimately mendacious superstructure.[7]

For Bloch, the utopian is conceived as a clear resistance to this impoverished way of reasoning and being:

> *The No could not be so strong if there were not among us, at the same time, a dangerous and battle-worthy Yes*; if at the same time, below this veiled life, below the nihilism of the modern age, a power [...] were not at the same time stirring [...]. Only this thinking, wishful dream brings about something real [...]. The urge to correspond with oneself draws soul into this dreadful world, into its unknowing, its error and its guilty conscience of its finality; everything has a utopian star in its blood.[8]

It is because of this resistance to instrumental reason that both Bloch and Benjamin were drawn to fissures in ordinary consciousness – the cracks, discontinuities, or what Peter Thompson has called the 'hole in the whole' – as sites of subversion where possibilities might emerge for something other than what is presented as the rock of a given and reified reality.[9] Thompson's characterisation of a 'hole in the whole' is reminiscent of Leonard Cohen's famous lyrics from his song 'Anthem': 'There's a crack in everything/That's how the light gets in'; and it is in these cracks that both Bloch and Benjamin, in different ways, search for utopia.

Bloch was motivated to look for places where there might still be an original wonder or amazement at what he called the 'primordial question'. He found this ongoing wonder in 'his preference for thinkers who regarded the world as an unrevealed mystery rather than a body of received laws and commandments', but also in his own search for the gold in the rubble, or the hole in the whole, of even the most mainstream and seemingly ideological cultures.[10]

Benjamin's search for the cracks and discontinuities led him in a slightly different direction. His primary interest was in the marginal or forgotten: in once-fashionable places and things, after their utility had dried up and a different truth thereby became visible in them; or in the child's encounter with a city or room, the magical enchantment of the child understood as an opposition to the 'cold estrangement of bourgeois adulthood'.[11]

For present purposes, this adds another guiding principle in our own hunt for utopia. As we have already seen in Chapter 3, utopia is that which resists 'what is', it is that which seeks to overcome material and spiritual alienation, and it is the germinative futural possibilities which are 'not yet' or 'still not'. But, as highlighted in this section, it is *also* helpfully conceived as that which resists instrumental reason: that which often emerges in the discontinuities and asides; that which is motivated by an original amazement at the inconstruable question, and does not automatically reduce everything to its finite place within the web of causality, utility and so-called explanation.

Notes

1 Walter Benjamin, 'The Lives of Students', in *Selected Writings*, ed. by Marcus Bullock and Michael W. Jennings, vol. 1 (Cambridge, MA: Harvard University Press, 1996), pp. 37–48 (p. 38).
2 Benjamin, 'The Lives of Students', *Selected Writings*, p. 46.
3 Benjamin, 'The Lives of Students', *Selected Writings*, p. 41.
4 Bloch, *The Spirit of Utopia*, p. 197 (my italics).
5 Benjamin, 'The Lives of Students', *Selected Writings*, pp. 43, 45.
6 Benjamin, 'The Lives of Students', *Selected Writings*, p. 46.
7 Bloch, *The Spirit of Utopia*, p. 167.
8 Bloch, *The Spirit of Utopia*, p. 171.
9 Peter Thompson, 'The Frankfurt School, Part 2: Negative Dialectics', *Guardian*, 1 April 2013, www.theguardian.com/commentisfree/2013/apr/01/negative-dialectics-frankfurt-school-adorno [accessed 2 July 2016].
10 Plaice, Neville, Plaice Stephen and Knight, Paul, 'Translator's Introduction', in Ernst Bloch, *The Principle of Hope*, 3 vols (Cambridge, MA: MIT Press, 1986), pp. xix–xxxiii (p. xxx).
11 Gilloch, p. 102.

4 The 'strange magic' of *Style Rookie* and *Rookiemag*

Figure 4.1 'My United States of Whatever'.

At a fleeting glance, the images in Figure 4.1 bring to mind a magazine cut-out or family photo album from an idealised 1950s America. The photo of the girl eating ice-cream evokes the familiarity of the mom-and-pop store, and her physical mimicry of the bear atop the store is suggestive of belonging, fitting in. In the last photo, she stands in front of the house, dressed like a dutiful housewife greeting guests at her home. The pink in her sunglasses and the fluffy coat that she holds match the pink framing of the house, again suggesting a unity of subject and setting: woman and home.

But a more careful look reveals greater complexity: both the era and intention of the images is hard to place. There are no actual clues as to the decade in which the photographs were shot, the background functioning as more of a generic suburban scene (to which any post-war decade might be related) than a purposively retrospective setting. The fashion the girl wears, while mostly 1950s inspired, includes pieces – such as the sparkly gold shirt – that are closer to 1970s disco-wear in style. Not only the fashions, but also the aesthetic of the images are not easily identifiable with a given era, or indeed location. The soft lighting and mostly pastel palette evoke a dreamy Californian aesthetic, although the lack of any actual Californian references suggest that it is more likely an inland suburban setting.

The mood, too, is awry. The contemplative expression of the Lolita-like girl expresses desire, longing, but also alienation. She does not stand on the porch of the pink house but outside the fence: in tune with the setting but also apart from it. Something – time or space – is out of joint. The bear in the ice-cream store brings to mind Jeff Koons' 1988 bears, implying both comfort and banality, and hinting at the layers of meaning that have accrued since the utopic 1950s imagery – smiley faces in sunny suburbia – that the picture initially brings to mind.

Despite these early indicators that something other than mere imitation or repetition is at play in these pictures, their palpable pastness is noteworthy. The girl in the photograph is in fact a fifteen year old in 2012. Her name is Tavi Gevinson and the photos (shot by her friend, photographer Petra Collins for *Oyster* magazine) appear on her blog, *Style Rookie*, under the suggestive blog post title 'My United States of Whatever'.[1] Tavi's writing and imagery is imbued with a particular kind of nostalgic sensibility that I will attempt to describe, elaborating how and where it is present in her work.[2] The primary question of this chapter is how to understand this nostalgic sensibility. In other words, I offer a reading of Tavi's mixed media output as a contemporary nostalgia text. I will ask: What is the nature of the nostalgia expressed in Tavi's writing and imagery? Is it regressive, a sign of cultural inertia, as a Jamesonian reading might suggest, or is there hope in it? If so, where do the utopian moments lie and how do they function?

Introducing Tavi

Tavi Gevinson was born in 1996 and raised in the Oak Park suburb of Chicago. Her father is a Jewish-American teacher of literature at Tavi's high school, and her mother (who was born in Norway and later converted to Judaism) is a weaver and part-time Hebrew instructor.[3] She has two elder sisters.

When Tavi was eleven years old, in 2007, she started the above-mentioned fashion blog, *Style Rookie*. The blog combined photographs of herself in varied and often outlandish outfits (shot in her neighbourhood) with mood boards, reflections on her life as a suburban teenager and commentary on both fashion trends and cultural works that inspired her. Tavi's style at this time was highly distinctive, incorporating 'eccentric granny' with teenage innocence and promise. Her writerly voice, too, was distinctive. Early on, *Style Rookie* attracted significant attention for its original aesthetic and intelligent commentary, and very quickly, if unexpectedly, Tavi was catapulted into a kind of twenty-first century fame.

Within two years, the blog was averaging 50,000 hits a day.[4] Tavi became a fashion sensation and was flown by large fashion companies to Paris, New York and Tokyo to attend their fashion shows. She was profiled by the *New Yorker* and *New York Times*, among other publications; and gave a talk at a TedX conference ('A Teen Just Trying to Figure It Out.').[5] By aged 13, Tavi had achieved a cult status not only as the kooky darling of the fashion world, but also as a kind of ambassador or representative for teenage girls. Her parents, who had anticipated an educated but ordinary middle-class American teenage experience for their children, were surprised. As her mother puts it in an interview, 'Did I think our children would be creative? Yes. Academic and like to travel? Yes. But.'[6]

Aged fifteen, Tavi had become somewhat disillusioned with the fashion world, writing on her blog that 'Fashion Week is weird. It is very high schooly' and that 'when I did look around a bit, no one looked happy and it was kind of a huge downer'. She turned 'to other places for a creative outlet and for inspiration'.[7] It was out of this that the second major stage of her teenage career emerged: in 2011 she founded *Rookiemag* (www.rookiemag.com), an online magazine for and by teenage girls, of which she was editor until she shut it down in 2018. The subject matter of *Rookiemag* is diverse, ranging from fashion to feminism, books to bullying and art to authenticity – all organised around monthly themes with evocative titles such as 'Enchantment', 'Forever', 'Longing', 'The Other' and 'The Great Unknown'. The magazine received one million views in its first week, and remained hugely popular among teenage girls, garnering considerable praise from adult readers as well.

Tavi has been described as a 'blogger turned online impresario', a 'pint-sized wunderkind' and as 'the oracle of girl-world'.[8] By her late teens, she was (and still is) a well-established media sensation and modern celebrity, with appeal to a range of literary, media and pop cultural figures. As one magazine article says, she has been 'adopted by the pop intelligentsia'.[9] She counts Stevie Nicks, Taylor Swift, Lorde, Courtney Love, Winona Ryder, Zadie Smith, Hilton Als, Karl Lagerfeld, Lena Dunham, Miranda July, Ira Glass, Jimmy Fallon, David Sedaris, Anna Wintour, Sofia Coppola and John Waters among her friends and supporters: a highly desirable clique of film-makers, authors, artists, actors, TV personalities, critics and musicians – all with a certain cultural capital, or 'cool' credential. Innocent and sophisticated, outsider and insider, suburban and urban, feminist and feminine, and equally comfortable Instagramming with her coterie of famous friends or rolling her eyes at her father and the daily perils of teenage life, Tavi sits at the threshold between celebrity and middle-class American normality.

Defining the text

I refer to the object of analysis in this chapter varyingly as a 'nostalgia text', Tavi's 'work', 'media output', 'writing and imagery' or 'oeuvre'. This is reflective of a struggle for adequate terminology to describe a relatively recent phenomenon: the blogger/writer/producer of media content whose varied outputs are viewed as her 'work'; an artistic or literary expression that her readers perceive as akin, in gravity, to more traditional forms of writing or creative arts. 'Writer' is an identity and writing a practice which are clearly central for Tavi – as she puts it, 'I consider writing my main discipline just because it's the only thing I do every day [...]. I think like an archivist, so being a writer feels more part of my personal identity than anything else.'[10] However, like other contemporaries such as Miranda July and Lena Dunham, the boundaries between writing and other kinds of artistic expression are not at all clear: she takes photographs, sings, puts together collages and mood boards of other people's work, makes videos and more. Even with regard to the classifier 'writer', it is not easy to define the kind of writer that Tavi is: blogger, critic, essayist, interviewer, memoirist – she is all of these. Although ostensibly writing about fashion, pop culture and feminism, Tavi's appeal is also personal – her life-story recorded and narrated in widely-followed pictures, blog-posts, journals and videos.

All of which makes defining Tavi's text somewhat difficult. There is reason to treat all of Tavi's self-produced media output as a single, amorphous whole. One can trace enough themes and intentions across the varied content to justify thinking of her work as an ongoing, evolving, project. However, for the purpose of clarity, I will focus my analysis on four primary texts: First, Tavi's blog, *Style Rookie* (2007–2015), for which she personally produced all of the content. Second, the independently run online magazine that Tavi founded and of which she was editor-in-chief, *Rookiemag* (2011–2018). (I will only consider the monthly editor's letters that Tavi writes for the magazine, as well as occasional articles that she contributes. I will not include writing or content produced by other *Rookiemag* staff writers or contributors.) Third, a talk given by Tavi for Sydney Opera House's 'Ideas At The House' festival in August 2013. Tavi's talk, entitled 'Tavi's Big Big World', was just under an hour long, and focused on the merits of fangirling. It is available to view on YouTube.[11] Fourth, her personal diaries. Tavi is an ardent diarist and, although these are not all available for public consumption, excerpts from her diaries are shown both in *Style Rookie* and as part of her talk at the 'Ideas At The House' festival. I may also occasionally refer to writing that Tavi has done for other publications (online or in print) but my primary focus will be on these four texts, in which I will explore and analyse the evident nostalgia via close reading and discursive analysis.

My specific focus is on the five-year period between 2009 and 2014, when Tavi was aged thirteen to eighteen. The reason for this is twofold. First, there is a particular *teenage* quality to her writing and interests, and indeed to the kind of nostalgia that her work exhibits, such that it makes sense to limit the object of analysis to content produced when she was a teenager, writing from her bedroom in her parents' suburban Illinois home. Second, Tavi's life, career and her personal philosophy have changed substantially since then. She has moved to New York and started acting professionally, as well as continuing with various writing projects. She has also started to question many of the ideas, as well as approaches to writing, blogging and journaling, that governed her thought as a teenager. My analysis should not therefore be read as a comment on Tavi's current or more recent work, which is notably different in tone from the work under consideration here.

Notes on a new nostalgia

In 1964 Susan Sontag published the now well-known essay 'Notes on "Camp"' in *The Partisan Review.* In it, she attempts to describe a sensibility known as 'camp' which was then relatively new. She begins by noting the difficulties involved in such an undertaking: 'A sensibility (as distinct from an idea) is one of the hardest things to talk about.'[12] And she goes on to explain why she adopts the form (numbered notes) that she does to write about camp:

> To snare a sensibility in words, especially one that is alive and powerful, one must be tentative and nimble. The form of jottings, rather than an essay (with its claim to linear, consecutive argument), seemed more appropriate for getting down something of this particular fugitive sensibility.[13]

My present attempt is akin to Sontag's in that I hope to describe a sensibility – in this case, a new kind of nostalgic sensibility – that is evident in Tavi's work (my present focus) but that is also more widespread. I will therefore adopt Sontag's method of jottings, or notes, both as a nod to Sontag's pioneering approaches to (then) contemporary culture, and because I share her sense that linear argument is not the best method for capturing the essence of something so amorphous. By jotting down different noticeable facets of the particular kind of nostalgia exhibited in Tavi's work, I hope to piece together a whole picture: to portray something of her sensibility, as well as that of the cultural moment of which she is a part.[14] In order to let Tavi's work also speak for itself, I have included, alongside these notes, what I call 'exhibits': examples of Tavi's writing or imagery that exhibit the particular qualities that I highlight in the notes. I also draw occasional parallels with other examples of contemporary cultural production.

1 The new nostalgic sensibility, like all forms of nostalgia, is attracted to the past.
2 The allure is not toward any particular point, or points, in the past, but toward a more general past, a kind of pastness. Images, objects, styles and cultures of the past are readily mixed together without any focused referent.

- *Exhibit A:* The images in Figure 4.1 in which Tavi mixes fashions from different eras. They bring to mind filmmaker David Lynch's 'blurring of visual references to 1950s and 1980s American suburbia' in *Blue Velvet*, which Laura Mulvey argues is used in order to create the impression of an 'amorphous temporality'.[15]
- *Exhibit B:* A *Guardian* article describing yearbooks that *Rookiemag* has published annually (their only in-print publication) also picks up on this mixing of temporal cues:

> Every page of the yearbook is covered with doodles and collages, a pinky, chewing-gummy lens through which you see their world. Illustrations and photos hark back to a time long, long before its authors were born, a mystical late 60s or tinted 90s, or somewhere in a dream.[16]

3 However, not all pastness is equal. The framework of reference for this cultural nostalgia is very recent, focused primarily (although not exclusively) on post-war America – from the 1950s up to the present day.

4 The setting for this kind of nostalgia aesthetic is suburbia. The oscillation of temporal and geographical markers evokes a generic suburban past, not a specific time or place, nor a specific memory.

5 The nostalgic sensibility is so intertwined with filmic references that it is hard to distinguish longing for an actual earlier suburban existence from longing for life as if in a movie. This is arguably reflective of the wider culture in which suburbia is as much an idea as a reality. Robert Beuka observes that from early on in the massive postwar development of suburban housing, TV helped to invest the emergent landscape with symbolic meaning: the expansion of suburbia entailed 'the construction of not only a new physical landscape, but new psychic and emotional landscapes as well'.[17] It is this imagined landscape that the new nostalgic sensibility evokes, and longs towards.

- *Exhibit C:* Tavi's description of her walk to school highlights how important films are in informing her own experience of suburban life: 'The line of houses reminds me of *American Beauty*, making the sameness endearing, and the hill by my school reminds me of the show *My So Called Life* because you've got your hill and your fence and your kid smoking and the field just below and then all the angst kind of just becomes bittersweet and like something to cherish.'[18]
- *Exhibit D:* Blog post, *Style Rookie*, 3 April 2013. Tavi writes,

> When I remember eighth grade, I recall scenes my mind illustrated while reading *Norwegian Wood* just as well as, and in some cases more vividly than, classmate interactions and walks to school. I spent a lot of freshman year analysing my close, personal relationships with Rayanne Graff and Laura Palmer.[19]

6 There is a dreamlike, kitsch aesthetic that accompanies the new nostalgic sensibility and the accordant representations of suburbia.

- *Exhibit E:* Tavi frequently uses the term strange magic to describe both her fascination with the small, seemingly ordinary or even phony details of suburban life; and to describe the particular aesthetic to which she is drawn.
- *Exhibit F:* Tavi has often written about how much she has been influenced by Sofia Coppola's *The Virgin Suicides*, a film that she appreciates in part because of its 'tacky' and 'cheesy' and 'dreamy' aesthetic, with its 'montages of the girls prancing around with unicorns, yellows and pinks and suburban skies'.[20] Likewise, she says of *Moonrise Kingdom* by Wes Anderson (a filmmaker well-known for his kitsch suburban aesthetic) that I 'cried about how the world will never look this beautiful'.[21]

7 The new nostalgia is divorced from actual memory, bringing to mind Appadurai's concept of 'ersatz nostalgia' or 'armchair nostalgia': nostalgia, that is, 'without lived experience or collective historical memory'.[22]

- *Exhibit G:* Blog post, *Style Rookie*, 27 March 2012. Tavi writes that 'images related to the '60s or the Southwest have always felt like old memories to me, even though I was obviously not alive in the '60s and when we visited my grandparents in California when I was little – my only experience in the Southwest – we spent the whole time inside their house'.[23]
- *Exhibit H:* Blog post, *Style Rookie*, 29 December 2011. Tavi writes about an abandoned building near her home that will not be torn down because an important law was debated there: 'It makes me nostalgic for something I've never really experienced which is, I guess, life without the internet, when those characters' records were some of the only access they had to a world outside changing families and classmates they'd known for years.'[24]

8 There is also nostalgia for the very recent past. As Coon observes, due to faster cultural turnover, 'instead of reminiscing about styles, behaviours and customs a hundred years in the past, we nostalgically remember cultural icons, catch phrases and electronic gadgets from less than five years ago'.[25] Indeed, we seem able to feel nostalgia for the present moment itself, which brings to mind Appadurai's discussion of a 'tension between nostalgia and fantasy where the present is represented as if it was already past'.[26]

- *Exhibit I: Rookiemag*'s call for submissions, written in 2013: 'we are big fans of the vintage, old-timey feel of static TV from 2002'.[27]
- *Exhibit J:* Blog post, *Style Rookie*, 3 April 2013. Tavi reflects on a recent day in a passage that is already imbued with a heightened consciousness of its passing:

My boyfriend and I take a tiny road trip during spring break. We skip stones on the beach, drink Coke out of glass bottles, and watch pink sunsets settle into night time. We walk along train tracks in the dark and stop to look at an old car behind a restaurant. I ask him to stand in front of it so I can take a Polaroid, the only picture I would have of him. I retrieve it from my bag once we're on the dull Midwestern highway, leaving for good. The photo got exposed in the streetlight and came out as a mess of brown and blue spots. In a panic, I rapidly relay the day's events in my mind, and jot down a few details to remember.[28]

9 There is a particular privileging of the 'teen' experience in real time, by teenagers, lending it in immediacy the rose-tinted glow that nostalgia used to provide only many years after the event.

- *Exhibit K:* Editor's letter for *Rookiemag*, December 2013. In this post, written when Tavi was still sixteen, she reflects on the teenage years (a period she names 'Forever') and what makes them special: '*Forever* is the state, exclusive to those between the ages of thirteen and seventeen, in which one feels both eternally invincible and permanently trapped.'[29]

10 The new nostalgic is self-consciously nostalgic. Nostalgia is something she enjoys, even courts.

- *Exhibit L:* Tavi frequently reflects on her own nostalgia, which she describes in one blog post (*Stylerookie*, 3 April 2013): 'when the act of remembering becomes more enjoyable than the event itself, conjuring feelings that are warm and fuzzy, but also *painful* in the best way'.[30]
- *Exhibit M:* Article on 'Moving Images: Shows and Films Imbued with Yearning', *Rookiemag*, June 2013. Tavi describes herself as 'someone who gets nostalgic for EVERYTHING: that water bottle wrapper, my old desk wallpaper, last Tuesday'.[31]

A Jamesonian reading

As discussed in Chapter 1, Fredric Jameson's 1984 diagnosis of the 'nostalgia mode' in American culture followed on from the 1970s nostalgia boom, and was one of a series of attempts around that time to understand and make sense of the rise in nostalgic cultural production. Grainge observes that most then-contemporary efforts understood the 'nostalgia boom' of the 1970s as a response to a sense of national crisis or malaise. For example, Fred Davis argued that nostalgia was tied to a sense of dislocation and Allison Graham attempted to explain it in terms of creative exhaustion. In contrast, Jameson (and, subsequently Grainge also) understood these contemporary expressions of nostalgia less as a 'mood' of longing or loss than a new 'mode' that developed around that time.[32]

Over thirty years on, it is easy to see why Jameson's analysis of the 'nostalgia mode' is still frequently called on to understand and explain nostalgia texts. His analysis is highly applicable in contemporary contexts. Indeed, the nostalgic sensibility exhibited in Tavi's work can, on one level, be read as exemplary of Jameson's 'nostalgia mode'.

The most obvious starting point for a Jamesonian reading of Tavi's output is the imagery that proliferates on *Style Rookie*. Reams and reams of images move fluidly between different historical moments and different contexts, with very little explanation. Sometimes the decade of different styles is referenced, but always in relation to the feeling or mood that the style evokes in the contemporary viewer, never in relation to the actual character of society or life at the time. For example, Tavi says of a collection of pictures in one blog post that 'they fit in so much with what I've been feeling for spring – the '50s/'60s jumble, the beehive, applying the bright pink/green/orange/sky blue colours I normally associate with Los Angeles to '50s shapes I normally associate with being home in a suburb'.[33] Sometimes the images Tavi uses are photographs or film stills from the past, and other times they are contemporary photographs (often her own) that replicate previous eras or film scenes in fashion or mood: there is little to distinguish between the two kinds. The imagistic imaginings of the past presented in Tavi's blog posts and mood boards could easily be accused of approaching the past, in Jameson's terms, 'through stylistic connotation, conveying "pastness" by the glossy qualities of the image, and "1930s-ness" or "1950s-ness" by the attributes of fashion', thus turning history into pastiche, a flat and random collage of dead styles.[34]

Admittedly, although Tavi discusses a wide range of topics on her blog, *Style Rookie* is, first and foremost, a fashion blog, meaning that 'style' is its legitimate focus. It therefore has less responsibility than, for example, a film set in the past, to convey historical meaning or to be historically accurate. It is also worth pointing out that Tavi is aware that the past she dreamily evokes on her blog and elsewhere it is not the whole picture. She is also aware of the role the internet plays in obscuring as well as promoting a certain kind of pastness. For example, in an article for *Rookiemag* titled 'Why the 21st Century Isn't All That Bad' she writes, with typical self-awareness:

> I know I'm not alone in imagining that being alive in the 1960s meant I would walk down a naturally yellow-tinted street, wave to my friends John, Paul, Ringo and George, hop into a van of friendly strangers, and drive across America until we found ourselves (and *ourselves*) in California, Joni Mitchell waiting for us at some canyon next to a lady who sells magical beads on a rug. Or that in the '80s, I would've gone to school with Molly Ringwald during the day, and gone to clubs with Debbie Harry and Stephen Sprouse at night. Tumblr is full of teenagers moping about being born in the wrong decade and YouTube is full of people complaining that the Little Monsters of today will never know music like it once was. Most of the time I feel convinced that the whole internet is devoted to trying to preserve, remember and figure out what life was like *without* the internet.
>
> But you know something? It's not all that bad, this here 21st Century.

Further on in the same article, she continues:

> I don't want to upset any adult readers who maintain that the '60s or '70s or what have you actually were that magical (hi Dad), but they sure weren't the version I've scraped together from old photographs and movies.[35]

In spite of this awareness, it is nonetheless the case that the tens, or hundreds, of thousands of teenage girls who read Tavi's blog both consume, and mimic on their own blogs and tumblrs, this stream of vague, dreamy pastness. Images of groups of teenagers bicycling along wide suburban roads or lounging on railway tracks, bored housewives in faded pastels or the sparkly-grubbiness of a gym prom dance abound on the blogs and tumblrs of 'Rookies' (Tavi's followers): an amorphous temporally and geographically blurred suburban dream, largely devoid of actual historical content.[36]

Writing of the growing popularity of old photographs, Raphael Samuel, as early as 1994, argues that the content of an image is often secondary to its colour or tone. Grainge concurs: 'the difference between a print of a 1906 football team and women demonstrating for suffrage will be of little consequence, so long as both are in sepia'.[37] Both statements ring true of today's nostalgic tumblr'ing of pastness. The use of imagery on *Style Rookie* and in other blogs also brings to mind, more intensely than he could have imagined, Siegfried Kracauer's 1927 essay *On Photography* in which he wrote of his fear that 'the flood of photos sweeps away memory's dam' and 'the blizzard of photographs betrays their indifference to what things mean'.[38]

As mentioned in Chapter 1, Grainge makes the point that, unlike the nostalgia 'mood' which is linked to longing and loss, the commodification and aestheticisation of nostalgia in the nostalgia 'mode' – a consumer appetite for pastness – is often simply a reflection of technical and economic developments. He explores the example of the Reagan-era launch of a new 24-hour cable TV station called the 'nostalgia network', which he argues should be explained not only in relation to the political agenda of the New Right, but also in the context of the massive explosion of cable television during the 1980s and the fast-growing market segment of baby-boomers: straightforward technical and consumer or market developments, in other words.[39] Likewise, it is hard not to perceive the proliferation of imagery from previous decades on *Style Rookie* and on other teen blogs as, at least in part, simply a function of the increased access to images and cultural trivia from the past that the internet makes possible. The nostalgia in *Style Rookie* seems thus to be to some extent a reflection of new technologies, rather than of new longings. This could be read as further evidence that Tavi's nostalgia is a classic example of the nostalgia 'mode': a taste for styles which connote pastness and can be (thanks to the internet) consumed or enjoyed in the present.

In addition to the ways in which she uses imagery on *Style Rookie*, Tavi's approach to memory could also be read as evidence of a nostalgia 'mode' operative in her work. Tavi is an attentive, almost obsessive, documenter and archivist. Both online in her blog and offline in countless diaries, Tavi collects and chronicles everything: memories of her own experiences, details from film scenes, lines from books, song lyrics and more. These diaristic collections are sorted and categorised according to Tavi's personal associations, an internal landscape of memory and points of connection. I will refer to

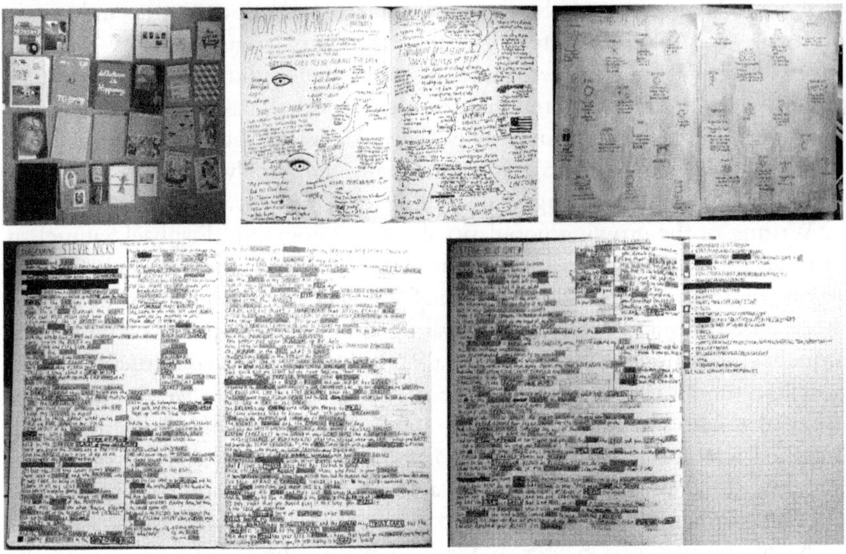

Figure 4.2 Tavi's notebooks.

Tavi's archivist practice and her 'collection' interchangeably: what I mean by both is the ongoing assemblage of memories and aesthetic details which she records, both in her many personal diaries and online in her blog.

It is notable that in Tavi's 'collection', memories of her own lived experiences, of watching films, and of listening to music or reading books, and moments from within those films or books, are chronicled alongside one another. Moreover, she does not seem to treat them as fundamentally different in kind. Filmic or literary references, and her felt responses to those cultural products (where and when she first watched, read or heard something and what it meant to her), are thus not only an important part of her diaristic archive, but also, one gets the feeling, of her lived experience. Her desire to document is something she has frequently written and spoken about, as in the following blog post, quoted at length:

> The habit that blog-keeping instilled in me of compulsively archiving *every single thing* only worsens. If I get behind in my journal, I spend hours wondering where to even start. I can't pay attention in class, only making scattered notes where there should be a timeline of the Industrial Revolution, listing all the details I need to get down properly as soon as I have time: The music we listened to in Claire's room, the old man I saw on my way to school, the view from my boyfriend's car when we sat in a 7-Eleven parking lot watching people walk in and trying to predict their purchases, along with a record of what each person looked like and what they bought. My hands tremble, relaxing only once everything has

> been sufficiently documented, each memory in my grasp, as if by putting them down on paper I can make them last forever.
>
> I develop my own form of sacred geometry to find the secret knots among these details and fit them into the rest of my journal. I go through one every two months or so, and for that period of time, coordinate it and all other parts of my life with a specific mood. My handwriting, my doodles, the clothes I wear, the books I read, the music I listen to, the movies I watch and the streets I walk down all match up. One goal of this is to create memories that are aesthetically pleasing and perfect and synesthetic, each element in place (and never repeated in another journal or memory, making its singular usage extra special) so that the nostalgia will feel *extra* good.[40]

Susan Stewart distinguishes between the souvenir and the collection.[41] The souvenir, she argues, is connected to memory: its point is to remind one of specific moments or places. The collection, by contrast, is connected to nothing but itself: it 'seeks a form of self-enclosure which is possible because of its ahistoricity'. The collection replaces history with classification, with order beyond the realm of temporality.[42] Tavi's attentive documentation confuses this neat distinction. Although its object is memory, it is closer in kind to Stewart's description of the collection than of the souvenir – more concerned, as evidenced in the preceding blog post, with a kind of coordination of moods, or 'sacred geometry', than with any tangible connection to specific events. Tavi could be described as a 'collector of memory' – whether her own memories or memories of things she has read about, watched or listened to – and her blogging practice as a testament to this collection. She is able to move so fluidly between the cultural and the personal because the focus is not on what happened or who it happened to or even whether it happened; it is on the feeling and aesthetic experience that something evokes, or evoked. What Stewart says of the collection is true of Tavi's blog, notebooks and personal archives: the ultimate referent is 'the interior of the self'.[43]

Tavi brings memory into a neat, ahistorical classification, and this neat, ahistorical, classified world is, for her, a source of nostalgia – as she says, one of the reasons for this collection of aesthetically pleasing memories is 'so that the nostalgia will feel *extra* good'. Elsewhere, Tavi also links her habit of hoarding memory to a desire to exacerbate the nostalgia:

> Now I've been combining my *aesthetic catalogs* with my diary, because it was too hard to carry around a million (or two) books at once and then it's like double the nostalgia and I'll feel really accomplished and complete when I look back and have these cohesive memories.[44]

Jameson's argument that cultural production is 'driven back inside a mental space' where 'it must trace our mental images of the past upon confining walls' could well be applied to Tavi, whose 'mental space' is an extraordinary memory collection, each memory or association brought in to her present order, within which she can 'enjoy' the nostalgia safely.[45]

A Blochian reading

Such a Jamesonian reading of Tavi's nostalgia is, in some ways, convincing. It understands Tavi's work as exemplary of the nostalgia 'mode': Tavi and her fellow Rookies as contented consumers of pastness. And one could stop there. But does this do justice to her rich imagination and longings, or to the chord that they clearly struck with readers of her blog? Is there not more at stake in these longings – in this dreamy, suburban nostalgia – than the consumption of a superficial and stylised past made possible primarily by the abundance of historical pop cultural imagery now available on the internet? I am reminded of Žižek's call to arms, in which he sees Bloch as a crucial ally: denunciation of ideology is not enough, one must also look for the authentic dreams, the utopian core.[46]

In my 'Jamesonian' reading, I highlighted Tavi's desire to revel in the nostalgia enabled by her 'memory collection' (by bringing the past into her own present order, she is better able to enjoy the nostalgia). However, there are other ways of reading Tavi's collecting practice, ones which I believe bring us closer to understanding the broad appeal of 'Tavi's world'.

Walter Benjamin, himself an avid collector of books, writes about the collector both in his famous essay 'Unpacking my Library: A Talk about Book Collecting' and in the less-well-known Convolute H, 'The Collector' in *The Arcades Project*. Aligning with Stewart's focus on the ahistoricity of the collection, Benjamin too recognises that the collector removes objects from history and divorces memories from their proper contexts:

> Thus represented, the things allow no mediating construction from out of 'large contexts.' The same method applies, in essence, to the consideration of great things from the past […]. We don't displace our being into theirs; they step into our life.[47]

However, Benjamin also highlights another very important aspect of the collector's work, which is that the collector adopts 'a relationship to objects which does not emphasise their functional, utilitarian value – that is, their usefulness – but studies and loves them as the scene, the stage, of their fate'.[48] In other words, 'The collector delights in evoking a world […] in which things are freed from the drudgery of being useful.'[49] For Benjamin, this desire to view things independently of their utilitarian value, to re-function them in entirely new contexts, is hugely important. It makes clear that the act of collecting can be fundamentally anti-utilitarian: the objects are celebrated as ends in themselves, and the collection has no use outside of its own parameters. As Hannah Arendt writes in her introduction to Benjamin's volume of essays, *Illuminations*, in this effort to redeem objects as no longer merely means to ends, but as having intrinsic worth, Benjamin 'could understand the collector's passion as an attitude akin to that of the revolutionary'.[50]

Benjamin initially wrote about the collecting of books, although as Arendt points out, in his later years he came to collect quotations with a perhaps even greater fervour, assembling 'a sort of surrealistic montage' which tried 'not to ruin everything with explanations that seek to provide a causal or systematic connection'.[51] What would Benjamin, whose *Arcades Project* is arguably the most impressive collection of quotes ever, have made of young Tavi's vast collection of memories, associations, aesthetic details and feelings?

I believe that Tavi's practice of collecting and archiving can likewise be read as an act of salvation, one that it is motivated by a strong resistance to instrumental reason and utilitarian logic. In her archival collections, she reassembles things according to a logic of an aesthetic and personal nature, rather than understanding them according to their objective use value. In Tavi's archival system, a 7-Eleven car park becomes not just a place designed to sell cheap consumer goods, but a red sign against blue sky, or the point at which one man passes by another every day. And it is logged in her memory-bank according to these designations, not according to its actual utilitarian function. Likewise, cheap, tacky paraphernalia are re-imagined as storehouses of dreams and longings, and so archived for posterity.

In the blog post cited previously – the one in which Tavi wistfully describes a recent road trip with her boyfriend and her felt urgency to jot down and remember the details ('In a panic, I rapidly relay the day's events in my mind, and jot down a few details to remember') – she concludes with the following sentence, a sentence that sums up poignantly the transformation of the world around her that she attempts via her archivist practice:

> The endless grey road with its yellow lights begins to feel less like a stretch of perpetual sameness, and more like an infinite sky filled with stars.[52]

If 'grey road' with 'yellow lights' is the utilitarian description of her surroundings, then there is a sense in which the act of recording her experiences and memories allows Tavi to aestheticise and in some way transform these surroundings; from within the ordinary she salvages something far more expansive ('infinite sky filled with stars').

In the ordinary details of ordinary life, and in the ordinary feelings of ordinary people, among what Bloch calls 'little daydreams', Tavi searches for wonder. Further evidence of this is found in the shrines that Tavi creates and photographs and which feature countless times on her blog (see Figure 4.3). The shrines, typical of a teenage girl's bedroom, are inspired at least in part by Sofia Coppola's representation of the bedroom of the Lisbon sisters in *The Virgin Suicides*. They combine cheap, tacky religious iconography with photographs

Figure 4.3 Bedroom shrines on *Style Rookie*.

of movie stars or boys that are crushed on; loved and overplayed records; favourite books and mixtapes; and little trinkets like bracelets or glittery diaries. It is significant that iPhones or laptops almost never appear in these visual scenes of longing – a testament, again, to the nostalgic tone of Tavi's collections. In these dreamily nostalgic assemblages of humdrum, kitsch paraphernalia and the minutiae of suburban life, Tavi sees hope.

The logic of Tavi's strange archive of personal memory, cultural memory and aesthetic details, is 'strange magic'. This is the term she uses most often to describe that which she wants to preserve, that by which she is fascinated. The way she has of talking about this 'strange magic' lends further credence to the idea that her work is a work of salvation. In particular, the religiosity of her lexicon in her descriptions of this 'strange magic' is striking.

In the talk she gives at the 'Ideas At The House' festival in Sydney, Tavi uses several different literary or filmic touchpoints to describe what she means by 'strange magic'. One of her reference points is the novel *Franny and Zooey* by J. D. Salinger, in which Tavi draws attention to the figure of the 'fat lady'. In the novel, when Franny and Zooey are young children who appear on radio shows, their older brother Seymour tells them to 'do it for the fat lady', an idea he has in his head of an ordinary woman sitting at home, really eager to hear the show. There's a passage at the end of the novel when Franny says to Zooey,

> There isn't anyone *any*where that isn't Seymour's fat lady. Don't you know that? Don't you know that goddam secret yet? And don't you know – *listen* to me, now – *don't you know who that Fat Lady really is*? … Ah, buddy, Ah buddy. It's Christ himself.[53]

Tavi interprets this passage as follows: 'The book seems to suggest that the highest being, the supreme calling, is a fan; Is the person who is open and wants to love things and wants to feel connected.'[54]

Another reference point that Tavi mentions is Chris Kraus's book, *I Love Dick*. The semi-autobiographical novel is about the character Chris's obsession with an academic named Dick, to whom she writes many unreturned love letters. In the end, the novel is more about the transformation that Chris undergoes via her love than about Dick, who is something of a straw man. Tavi reads this as confirming, as with the fat lady, that openness to love, the ability to love, and to wonder at the world, is of the greatest importance. Moreover, just as the fat lady is designated 'Christ himself', so too Tavi quotes the lines when Kraus invokes Jesus to describe Dick, or more accurately the loving, or longing, that Chris feels for Dick:

> You're shrunk and bottled in a glass jar, you're a portable saint. Knowing you's like knowing Jesus. There are billions of us and only one of you so I don't expect much from you personally. There are no answers to my life. But I'm touched by you and fulfilled just by believing.[55]

'Strange magic,' Tavi says, 'has now become shorthand for my own wonder, or the fat lady, or Dick.'[56] She also mentions light: 'I think I'm most attracted to light because of what it means in most literature. It symbolises some kind of truth or faith, like some "strange magic".'[57] And comments elsewhere: 'being a teenager can be lonely as hell, and in such moments, two things seem somewhat holy: Friendship and fandom. They create a sort of *strange magic*, if you will'.[58] 'Wonder', 'truth', 'light', 'faith', 'magic', 'holy': these are serious adjectives indeed to apply to the chintzy objects of teenage girls' bedrooms or the little longings that they signify. They are big questions, too, to seek answers for amid such day-to-day rubble. To see the 'fat lady' as 'Christ himself' is surely an act of salvation: a transformation of the ordinary into the ultimate, a beatification of the banal. Likewise, Tavi's insatiable search for the 'strange magic' or wonder (or, even, 'Christ himself') in what might otherwise be conceived (and experienced) as a drab suburban existence is an attempted act of salvation; in her dreamy, chintzy, nostalgic, collection of 'strange magic' she redeems things from their fate as purely useful, mundane objects and in doing so redeems her own suburban existence from the depressing picture of it that is more often painted.

Several critics have observed that representation of suburbia in American literature and culture has oscillated between the utopian and dystopian. Beuka notes that there is a fundamental debate over the nature of the suburbs as either 'utopian models of community or dystopian landscapes of dispiriting homogeneity'.[59] The image of sunny utopia that emerged in the 1950s, particularly in televisual and filmic representation, presented the suburbs as being 'miraculously free from the serious problems of the rest of the world'.[60] Almost contemporaneously, however, critical voices began to emerge, as in David Reisman's *The Lonely Crowd* (1950), Betty Friedan's *The Feminine Mystique* (1963), John Keats' *The Crack in the Picture Window* (1956) and Richard Yates' *Revolutionary Road* (1961), which depicted a darker, more dystopian, suburbia.

Traces of the uncomplicatedly happy utopian ideal continue up to the present day to inform literary and cultural representations of suburbia, as in the sitcom *Modern Family* (2009–) and D. J. Waldie's memoir, *Holy Land: A Suburban Memoir* (2005), but dystopian representations are far more common. As Beuka points out, American fictions and films from the past half century have depicted suburbia as 'a consistently negative environment', 'a contrived, dispiriting and alienating place'.[61] The questions usually asked of the suburban setting, as Murphy points out, are: 'What has been happening to these people? What is missing? What is so terribly wrong in this pretty green community?'[62] Indeed, each of these extremes is by now so familiar that, as Beuka points out, familiar representations of suburbia almost instantly evoke both the dream and the nightmare:

> Mere mention of the word 'suburbia,' after all, will call to mind for most Americans a familiar string of images – the grid of identical houses on identical lots, the smoking barbeque, the swimming pool – loaded signifiers that, taken together, connote both the 'American dream' as it was promulgated by and celebrated in popular culture in the post-war years and that dream's inverse: the vision of a homogenized, soulless, plastic landscape of tepid conformity, an alienating 'noplace.'[63]

By contrast, Tavi's representation of suburbia is neither the straightforwardly utopian imagery of the 1950s nor the more common dystopia that has dominated film and fiction since then. Her insistent and urgent pursuit of the magical in the mundane is, in its own way, startlingly reminiscent of Bloch, who, as Adorno puts it, 'hunts around in kitsch for the transcendence that is blocked by the immanence of culture'.[64] In her own hunt through kitsch Tavi herself is a kind of miner, searching for traces of all-too-human hope, of 'strange magic', which might otherwise be disregarded or overlooked. In this way, she achieves via her recording of these 'strange magic' details a kind of transformation of her environment and others like it.[65] Suburbia, far from being a false ideal or the 'homogenized, soulless, plastic landscape' that Beuka describes, becomes a place of dreams, of longing, of hope: often frustrated, thwarted, or misdirected admittedly, but there in plain sight for the eyes that are willing to look.

For Benjamin, the collector's process is one of renewal. He writes in his essay 'Unpacking my Library':

> [...] to a true collector the acquisition of an old book is its rebirth. This is the childlike element which in a collector mingles with the element of old age. For children can accomplish the renewal of existence in a hundred unfailing ways [...]. To renew the old world – that is the collector's deepest desire.[66]

Tavi sees her own practice similarly. She argues that 'reflecting and archiving is not the same thing as dwelling in the past. It is not anti-living, but a part of life, even a crucial one'.[67] Quoting Virginia Woolf's question, 'How can we combine old words in new orders so that they survive, so that they create beauty, so that they tell the truth?', Tavi suggests that in her work she is trying 'to create new order out of the things I love about the past'.[68] This perspective, the perspective of a collector about her own collection, aligns well with Benjamin's understanding of his own passion for book collecting. Although the objects may be of the past, in the collector's practice Benjamin reads a very Blochian sense of anticipation:

> [...] it is certainly not an elegiac mood but, rather, one of anticipation – which these books arouse in a genuine collector.[69]

Likewise, I have argued that, despite the palpable 'pastness' of Tavi's virtual and diaristic archive, it is not an elegiac mood but rather one of anticipation which it arouses, not only in the collector herself but in the many enthusiasts who follow her.

Metamodern nostalgia

The term metamodernism came to prominence in 2010 in the work of Timotheus Vermeulen and Robin Van Den Akker.[70] They are among several cultural theorists who assume the end of postmodernism as a cultural dominant and are battling it out to describe the terrain that lies beyond post-modernism (competing theories include Bourriard's Altermodernism, Kirby's Digimodernism, Boym's Off-modernism and Lipovetsky's Hypermodernity).[71] Vermeulen and Van Den Akker use the term metamodern to describe the contours of a discourse, or 'structure of feeling' that they argue emerged in the wake of postmodernism. Citing numerous examples in art, film and architecture, they argue that the postmodern strategies of irony and pastiche (typified in the visual and plastic arts by, for example, Jeff Koons, Jake and Dinos Chapman and Damien Hirst) are being abandoned and that a new narrative of longing – for belief, for sincerity, for a better future – is increasingly evident.[72]

Vermeulen and Van Den Akker briefly cite the possible causes of the perceived shift in approach: the 'threefold threat' of the credit crunch, a collapsed centre and climate change. However, their focus is not on explaining the context or cause of this new structure of feeling, but on describing it, and pointing to its existence in contemporary culture. The authors use the prefix 'meta', drawing on the term metaxy (*μεταξύ*) as used in Plato's symposium, to designate oscillation.[73] The emerging metamodern affect that they describe is not a naïve hope. It is characterised by an oscillation between a 'modern' desire for *sens* and a postmodern doubt about the sense of it all – between, in other words, 'a typically modern commitment and a markedly postmodern detachment'.[74] They argue that if the modern outlook vis-à-vis idealism is 'fanatic and/or naïve', and the postmodern mindset 'apathetic and/or skeptic', then the current generation ('for it is, very much so, an attitude tied to a generation') chooses to act sincerely in spite of its awareness of the problem of doing so. The attitude of this generation 'can be conceived of as a kind of informed naivety, a pragmatic idealism'.[75]

Another common term associated with metamodernism (and indeed with the post-postmodernism debate) is 'New Sincerity', made popular in the 1990s by David Foster Wallace.[76] Wallace focused particularly on the prevalence of irony in postmodern culture, and conceived the New Sincerity as a fatigue with and desire to move beyond irony:

> And make no mistake: irony tyrannizes us […]. All U.S. irony is based on an implicit, 'I don't really mean what I am saying.' So what *does* irony as a cultural norm mean to say? That it's impossible to mean what you say? That maybe it's too bad it's impossible, but wake up and smell the coffee already? Most likely, I think today's irony ends up saying: 'How totally *banal* of you to ask what I really mean.'[77]

My goal here is not to evaluate how widespread metamodernism is, nor even the validity of its initial premise – 'the death of postmodernism'. If Jameson conceived postmodernism as the cultural dominant of late capitalism, there remains a debate as to whether metamodernism is the new cultural dominant or just a strain within a still predominantly postmodern culture, but one need not enter into this debate in order to observe that the 'structure of feeling' described by Vermeulen and Van Den Akker is undeniably evident in Tavi's work. In exploring Tavi's 'metamodern' affect I hope to find further clues to understanding the particularities of the nostalgic sensibility exhibited in her media output.

Tavi is on the one hand highly conscious of the difficulty of straight-forward sincerity, and seems to exhibit an almost inherited postmodern knowingness right from her earliest posts, as an eleven year old. She prefaces every opinion or piece of advice with a side-acknowledgement that this might appear 'cheesy', or 'preachy' (adjectives which serve to distance her in some way from what she is saying); she is ever wary of appearing naïve. At the same time, however, she longs for the possibility of being sincere, authentic, and genuinely enthusiastic. She is torn, it seems, between two poles.

For example, the following blog post appeared in handwriting (as opposed to typed), in the form of a photograph of her diary, but not without considerable explanation:

> I don't know why I'm handwriting this post but I had my little book with me and it kind of looks nice. Plus I want you to know that I am just like, the total girl next door, just like, you know, a regular teen, as indicated by my loopy handwriting and quirky doodles […] (I can't get over how much more pretentious everything looks in HANDWRITING, but I guess I would rather look overly pleased with writing

> about stuff I like than apathetic when I'm actually totally excited ... plus I hate to think that even HANDWRITING now looks insincere because of its constant presence in women's magazines and fake handwriting fonts ... WHAT A WORLD, am I right?).[78]

In her talk at the Sydney Opera House she again exhibits this oscillation between a desire for *sens* and doubt about its possibility:

> Lately I've been thinking a lot about originality and authenticity and that's really what's got me all blocked up when I tried to work on this [...] actually it's at a point now where doing what I'm doing now, calling myself out for being clichéd, is clichéd, so I don't know what I'm supposed to do.[79]

Her amusing description of struggling to find her own voice in the internet age, in which it feels like everything has already been written or created and archived online, gives voice also to her perception of the precariousness or near-impossibility of authentic creativity:

> I'd continually read over what I would have so far and felt like I was reading a teenager's diary on a sitcom written by a middle aged props master.[80]

The problem facing Tavi is a problem articulated by Jameson many years previously:

> this very triviality of daily life in late capitalism is itself the desperate situation against which all the formal solutions, the strategies and subterfuges, of high culture as well of mass culture, emerge: how to project the illusion that things still happen, that events exist, that there are still stories to tell, in a situation in which the uniqueness and the irrevocability of private destinies and of individuality itself seem to have evaporated?[81]

However, and as distinct from Jameson's characterisation of the postmodern problematic, Tavi – in spite of her hyper-awareness of the difficulty of any authentic act of creativity or even of genuine enthusiasm – is still insistent that it be possible. This insistence is evident, for example, in Tavi's ardent defence of fangirling. She says of herself: 'all I really want to do is make fan-art. I want to be a professional fangirl'.[82] Her talk at the Sydney Opera House is essentially a defence of fangirling: as the references to both *I Love Dick* and the 'fat lady' in *Franny and Zooey* show, Tavi is resolute about the importance of loving things and sees this as almost more important than the question of whether the thing you love is a worthy object. It doesn't matter that Dick does not love Chris back; all that matters is what it means to Chris to have loved. Likewise, Tavi argues that you shouldn't wince when something you once loved now appears banal, because 'the personal truth you once saw exists not in a thorough analysis of the words on the page but in the moment in which you felt like you'd received a gift'.[83] This is evident, too, in the teenage shrines that Tavi lovingly recreates: tacky Jesus figurines sit alongside pictures of smouldering heartthrobs and overplayed records, all of them a testament to longing rather than particular objects worthy of worship (in these scenes Jesus is effectively interchangeable with Keanu Reeves).

Of course, this 'philosophy of fangirling' is not unproblematic. Enthusiasm as a corrective to too much, or too serious critique, as for example Sontag advocated in her famous 'Notes on "camp"' essay in the Sixties, is different to enthusiasm as a catch-all replacement of critique – one which negates any objective distinctions. While Sontag defended the need for the kind of celebration entailed in camp taste in the Sixties, arguing that it 'is a kind of love for human nature', one which 'relishes, rather than judges' she later, in the Nineties, qualified that defence in light of a cultural direction she hadn't quite anticipated: 'that seriousness itself was in the early stages of losing credibility'.[84] Although Sontag's change of heart was lamented by some – Ellen Willis, for example, writes 'I imagine a smiling Theodor Adorno welcoming Sontag home. […] Sontag, it seemed, had ended up, well, a curmudgeon' – there is also weight in Sontag's view that what had once required a defence was now uncritically, blithely, accepted.[85]

However, what Sontag does not mention, but is evident in Tavi's defence of fangirling, is that not only seriousness, but sincerity and enthusiasm, were themselves under attack – so much so that a thirteen-year-old girl feels compelled to re-claim the right to sincerely enjoy things, without the disruption of ever-present irony. If we follow in Sontag's footsteps and read Tavi's 'philosophy of fangirling' dialectically – asking not whether it is a good or bad way of approaching things, but what it is trying to do vis-à-vis its particular moment – we will see that her embrace of sincere enthusiasm can only be properly understood in relation to the difficulty that her generation experiences in doing just that. 'What I'm interested in,' Tavi says of fangirls, 'is the enthusiasm, the refusal to try and act cool and disaffected.'[86] This, surely, is a rallying cry against the generation steeped in ironic posturing that preceded hers. Are fangirls like Tavi and her 'rookies' the 'literary rebels' that Wallace foresaw in 1993?

> The next real literary 'rebels' in this country might well emerge as some weird bunch of *anti*-rebels, born oglers who dare somehow to back away from ironic watching, who have the childish gall actually to endorse and instantiate single-entendre principles. Who treat plain old untrendy human troubles and emotions in U.S. life with reverence and conviction. Who eschew self-consciousness and hip fatigue. These anti-rebels would be outdated, of course, before they even started. Dead on the page. Too sincere. Clearly repressed. Backward, quaint, naive, anachronistic. Maybe that'll be the point. Maybe that's why they'll be the next real rebels. Real

> rebels, as far as I can see, risk disapproval. The old postmodern insurgents risked the gasp and squeal: shock, disgust, outrage, censorship, accusations of socialism, anarchism, nihilism. Today's risks are different. The new rebels might be artists willing to risk the yawn, the rolled eyes, the cool smile, the nudged ribs, the parody of gifted ironists, the 'Oh how *banal!*'[87]

The oscillation between modern longing and postmodern doubt also informs Tavi's nostalgia. On the one hand, she knows that, as mentioned previously, 'the 60s or 70s [...] sure weren't the version I've scraped together from old photographs and movies' and that the 60s didn't actually mean one would 'walk down a naturally yellow-tinted street, wave to my friends John, Paul, Ringo and George, hop into a van of friendly strangers and drive across America'.[88] But on the other hand, she is endlessly fascinated by this world of pastness which seems somehow to contain within it some magic (or 'strange magic') ingredient. It is not incidental that she coins the term strange magic after the song 'Strange Magic' by the band Electric Light Orchestra, which she describes as 'the perfect sad *earnest* 1970s suburban song'.[89] The palpable allure of being earnest is clear, but it is necessarily located in the 1970s: earnestness is not of Tavi's moment, at least not un-self-conscious earnestness.

In particular, it seems as though an important underlying thread of Tavi's nostalgia is the desire for life before the internet. Although she only rarely mentions this ('It makes me nostalgic for something I've never really experienced, which is, I guess, life before the internet') and is often complimentary about the positive potential of the internet, this desire is implicit in the representations that she chooses.[90] In the endless imagery on *Style Rookie*, there are rarely any visual references to recent (post-internet) technology: iPhones, laptops, the internet. Instead, a viewer with no context might think from viewing these images that the internet had not been invented (save, of course, for the fact that she was viewing the images on the internet). There is a fascination in Tavi's blog with a world that seems somehow simpler, more authentic, or perhaps more to the point, un-mediated.

Baudrillard famously understood nostalgia culture as an attempt to fill a void, the 'desert of the real' in postmodern culture. He writes that 'when the real is no longer what it was, nostalgia assumes its full meaning'.[91] If in 1981 Baudrillard had already described this 'desert of the real', a world of simulacra which 'threatens the difference between "true" and "false", the "real" and the "imaginary"', he surely would have seen Tavi as the poster-girl for the world of simulation.[92] After all, she is a true child of the internet, who does not differentiate between memories of her own and memories of film-scenes, and whose experience of her suburban neighbourhood is so mediated by filmic references that there is no 'real' or first-hand experience to access. And yet she is endlessly drawn to an imagined 'before', where the lost possibilities – sincerity, enthusiasm, authenticity, belief, truth – seem, somehow, to lurk. If we understand nostalgia along the lines developed in Chapter 3, then it might be seen not only as an attempt to fill a void but more purposefully (if not consciously) as motivated by ghosts which 'continue to "insist"' and which aim to re-awaken forgotten potentials in the present moment. Tavi's nostalgia for the Real (and the lost potentialities it entails: authenticity, belief, truth) can thus be understood, within the framework of her metamodern dilemma, as the stirrings of the ghost of the Real, the unresolved struggle of the Real to find new ways of existing in a highly mediated postmodern culture.

The profound longing exhibited in Tavi's work – for authenticity, sincerity, belief – can be read most simply, then, as a welcome corrective to a preceding generation too steeped in the postmodern strategies of irony and pastiche. But perhaps it can also be read on a grander scale. It is worth at least raising the question: If 'God is Dead', as he has supposedly been for many years, then why is this young teenager searching for him within the day-to-day paraphernalia of suburban kitsch? What if Tavi is not only salvaging 'the fat lady' by making her 'Christ himself' (that is, salvaging her ordinary suburban existence from the impoverished instrumentality that so often defines it, by hunting for the magical, or mystical, in the mundane), but is also trying to salvage 'Christ himself' (the lost possibilities of truth or belief) by resurrecting him in 'the fat lady'?

To conclude this analysis, I want to propose that the nostalgia evident in Tavi's practice is best designated as 'metamodern' nostalgia. In Chapter 1 I discussed the evolution of the term nostalgia from 'modern nostalgia' or the nostalgia 'mood' (an intense, wistful longing for time gone by characteristic, for example, of Proust) to 'postmodern nostalgia' or the nostalgia 'mode' (a depthless consumer mode, or way of relating to certain styles). What is interesting is that, as explored, Tavi's dreamy, temporally blurred, suburban nostalgia appears to exhibit aspects of both the nostalgia 'mode' and the deeper longings and tussles often associated with the nostalgia 'mood' (albeit in a more contemporary context and reflective of the particular tensions of the simulacral internet-dominated world that Tavi inhabits). One of the identifying characteristics of the nostalgia 'mode' is its total absence of loss or yearning (it is, as Higson puts it, 'surprisingly sweet and not at all bitter'[93]); but by contrast with this characterisation of contemporary nostalgia culture as merely light-hearted consumerism there is an observable urgency to Tavi's memory practice. As she says in the already quoted blog post, 'My hands tremble, relaxing only once everything has been sufficiently documented, each memory in my grasp, as if by putting them down on paper I can make them last forever.'[94] She does not simply want to record her memories and observations to be later enjoyed; she also fears losing them. She is driven by a kind of existential urgency. A new term, metamodern nostalgia, seems best to capture the particularities of this complex new nostalgia, which oscillates between an easy and enjoyable consumption of past styles, and deeper longings for wonder, sincerity, authenticity and belief.

Notes

1 Tavi Gevinson, 'My United States of Whatever', *Style Rookie*, 22 June 2012, www.thestylerookie.com/search?updated-max=2012-09-16T11:38:00-05:00&max-results=1&reverse-paginate=true&start=3&by-date=false [accessed 4 July 2013].
2 Since Tavi Gevinson is widely known by her first name I will refer to her as Tavi throughout this chapter, a departure from the conventional academic practice of referring to people by their surnames.
3 Lizzie Widdicombe, 'Tavi Says: Fashion Dictates from a Fourteen-Year-Old', *The New Yorker*, 20 September 2010, www.newyorker.com/magazine/2010/09/20/tavi-says [accessed 5 August 2014].
4 Katrina Onstad, 'Is Tavi Gevinson Girl Power's Last Chance?', *The Globe and Mail*, 13 September 2012, www.theglobeandmail.com/life/is-tavi-gevinson-girl-powers-last-chance/article4543738/ [accessed 16 August 2014].
5 Widdicombe, 'Tavi Says'; Emily Witt, 'Tavi Forever', *T Magazine: New York Times Style Magazine*, 6 June 2014, www.nytimes.com/2014/06/06/t-magazine/tavi-gevinson-on-rookie-magazine-and-growing-up.html?_r=0 [accessed 5 August 2014]; Tavi Gevinson, 'A Teen Just Trying to Figure It Out', YouTube, 9 April 2012, www.youtube.com/watch?v=6osiBvQ-RRg [accessed 4 October 2014].
6 Mary Schmich, 'Being Tavi's Mom: What's It Like to Be The Mother of a Teen Sensation?', *Chicago Tribune*, 6 April 2014, http://articles.chicagotribune.com/2014-04-06/news/ct-tavi-gevinson-schmich-met-0406-20140406_1_tavi-gevinson-berit-engen-steve-gevinson [accessed 4 October 2014].
7 Tavi Gevinson, 'Untitled Blog Post', *Style Rookie*, 27 March 2011, www.thestylerookie.com/2011/03/i-feel-like-photo-to-accompany-this.html [accessed 4 July 2013].
8 Michael Schulman, 'The Oracle of Girl World', *New York Times*, 27 July 2012, www.nytimes.com/2012/07/29/fashion/tavi-gevinson-the-oracle-of-girl-world.html [accessed 4 August 2014]; Hannah Baxter, 'Tavi Gevinson Reminds Us Why She's The World's Coolest Teenager', *Bust*, http://bust.com/general/12826-tavi-gevinson-reminds-us-why-she-is-the-worlds-coolest-teenager.html [accessed 4 August 2014].
9 Amy Larocca, 'At 18, Tavi Gevinson is a Fashion Veteran – and a Broadway Rookie', *The Cut*, 10 August 2014, http://nymag.com/thecut/2014/08/tavi-gevinson-from-fashion-to-broadway.html [accessed 10 August 2014].
10 Thea Liberty Nichols, 'The Only Thing I Do Every Day: Tavi Gevinson', *Art 21 Magazine*, 22 April 2013, http://blog.art21.org/2013/04/22/the-only-thing-i-do-everyday-tavi-gevinson/#.WBi_JdCLTnB [accessed 5 August 2014].
11 Tavi Gevinson, 'Idea's At The House: Tavi's Big Big World', YouTube, 26 August 2013, www.youtube.com/watch?v=6osiBvQ-RRg [accessed 4 October 2014].
12 Sontag, *Against Interpretation*, p. 275.
13 Sontag, *Against Interpretation*, p. 276.
14 The only other effort I know of to describe a kind of contemporary nostalgia in note form, presumably also inspired by Sontag, is by the New York based artist Peter Halley, 'Notes on Nostalgia', *New Observations*, 28 (1985), www.peterhalley.com/ARTISTS/PETER.HALLEY/NOTES.ON.NOSTALGIA.FR2.htm [accessed 12 January 2016]. The kind of nostalgia he describes is highly reminiscent of the 'nostalgia mode' described by Jameson in 'Postmodernism', published in *The New Left Review* only a year earlier. For example, Halley's note 4 clearly reflects Jameson's idea that the nostalgia mode reduces history to a series of styles that can be consumed in the present:

4. History becomes entertainment and entertainment becomes history. There are ITV docu-dramas, and there are the Academy Awards. There is 60 Minutes, and there is the Baseball Hall of Fame. There is the traditional Christian calendar year 1984, and there is Super Bowl XVIII.

The sensibility I here describe in Tavi's work (30 years after Halley's piece) has some similarities, but also different nuances and inflections to the kind of nostalgia to which Halley attempts to give voice.

15 Martin Dines, 'Suburban Gothic and the Ethnic Uncanny in Jeffrey Eugenides' *The Virgin Suicides*', *Journal of American Studies*, 46.4 (2012), 959–975 (p. 963); Mulvey, Laura, 'The Pre-Oedipal Father: The Gothicism of *Blue Velvet*', in *Modern Gothic: A Reader*, ed. by Allan Lloyd Smith and Victor Sage (Manchester: Manchester University Press, 1996), pp. 36–65 (p. 55).

16 Eva Wiseman, 'Tavi Gevinson: The Fashion Blogger Becoming the Voice of a Generation', *Guardian*, 9 December 2012, www.theguardian.com/fashion/2012/dec/09/tavi-gevinson-fashion-blogger [accessed 5 August 2014].

17 Robert Beuka, *SuburbiaNation: Reading Suburban Landscape in Twentieth-Century American Fiction and Film* (New York: Palgrave Macmillan, 2004), p. 10; David R. Coon also observes that Suburbia exists in our cultural imagination as much as in our landscapes (p. 10).

18 Gevinson, 'Idea's At The House: Tavi's Big Big World'.

19 Tavi Gevinson, 'Five Years', *Style Rookie*, 3 April 2013, www.thestylerookie.com/search?updated-max=2013-12-03T00:51:00-06:00&max-results=1&start=2&by-date=false [accessed 4 July 2014] (Note: Rayanne Graf is a character from the TV show *My So Called Life* and Laura Palmer a character from the TV show *Twin Peaks*).

20 Tavi Gevinson, 'Teenage Disconnect and *The Virgin Suicides*', npr, 26 December 2012, www.npr.org/2012/12/26/163570963/teenage-disconnect-and-the-virgin-suicides [accessed 4 October 2014].

21 Tavi Gevinson, 'A Hodgepodge, If You Will', *Style Rookie*, 5 January 2012, www.thestylerookie.com/2012/01/hodgepodge-if-you-will.html [accessed 4 July 2014].

22 Appadurai, p. 78.

23 Tavi Gevinson, 'There Are Holes in the Universe', *Style Rookie*, 27 March 2012, www.thestylerookie.com/search?updated-max=2012-04-01T16:08:00-05:00&max-results=1&start=8&by-date=false [accessed 4 July 2014].

24 Tavi Gevinson, 'Strange Magic', *Style Rookie*, 29 December 2011, www.thestylerookie.com/search?updated-max=2012-04-01T16:08:00-05:00&max-results=1&start=8&by-date=false [accessed 4 July 2014].

25 Coon, p. 39.

26 Appadurai, p. 83.

27 *Rookiemag* submissions page, http://rookiemag.com/you [accessed 2 June 2013].

28 Gevinson, 'Five Years'.

29 Tavi Gevinson, 'Editor's Letter – December 2013: Forever', *Rookiemag*, 2 December 2013, www.rookiemag.com/2013/12/editors-letter-26/ [accessed 3 December 2013].

30 Gevinson, 'Five Years'.

31 Tavi Gevinson, 'Moving Images: Shows and Films Imbued with Yearning', *Rookiemag.com*, June 2013, Www.rookiemag.com/2013/06/movietv-rex/ [accessed 4 July 2013].

32 Grainge, p. 42.

33 Tavi Gevinson, 'Sixteen and Spring', *Style Rookie*, 3 May 2012, www.thestylerookie.com/search?updated-max=2012-05-20T09:28:00-05:00&max-results=1&start=5&by-date=false [accessed 4 July 2013].
34 Jameson, *Postmodernism* (1992), p. 19.
35 Tavi Gevinson, 'Why The 21st Century Isn't All That Bad', *Rookiemag*, 28 October 2011, www.rookiemag.com/2011/10/21st-century-not-all-that-bad/ [accessed 4 July 2013].
36 The same could not be said of *Rookiemag*, which often runs articles with an explicit focus on, or reference to, various historical events and figures.
37 Grainge, p. 41.
38 Kracauer, p. 51.
39 Grainge, pp. 47–49.
40 Gevinson, 'Five Years'.
41 Susan Stewart, *On Longing: Narratives of the Miniature, the Gigantic, the Souvenir, the Collection* (Durham, NC: Duke University Press, 1993), p. 23.
42 Stewart, *On Longing*, p. 23.
43 Stewart, *On Longing*, p. 158.
44 Tavi Gevinson, 'Technicolour', *Style Rookie*, 1 February 2012, www.thestylerookie.com/search?updated-max=2012-02-03T19:22:00-06:00&max-results=1&start=18&by-date=false [accessed 4 July 2014].
45 Jameson, *Postmodernism* (1992), p. 25.
46 Žižek, Preface, *The Privatization of Hope*, p. xix.
47 Walter Benjamin, *The Arcades Project*, trans. by Howard Eiland and Kevin McLaughlin (Cambridge, MA: Harvard University Press, 2002), p. 206.
48 Walter Benjamin, 'Unpacking My Library', in *Illuminations: Essays and Reflections*, trans. by Harry Zohn (New York: Schocken Books, 1969), pp. 59–69 (p. 60).
49 Benjamin, *The Arcades Project*, p. 19.
50 Arendt, p. 42.
51 Arendt, p. 47.
52 Gevinson, 'Five Years'.
53 J. D. Salinger, *Franny and Zooey* (London: Penguin, 1994), p. 150.
54 Gevinson, 'Idea's At The House: Tavi's Big Big World'.
55 Chris Kraus, *I Love Dick* (Los Angeles: Semiotext(e), 2006), p. 104.
56 Gevinson, 'Idea's At The House: Tavi's Big Big World'.
57 Gevinson, 'Idea's At The House: Tavi's Big Big World'.
58 Tavi Gevinson and Petra Collins, 'Strange Magic', *Rookiemag*, 8 October 2012, www.rookiemag.com/2012/08/strange-magic/ [accessed 4 July 2013].
59 Beuka, p. 7.
60 Coon, p. 13.
61 Beuka, p. 228.
62 Bernice Murphy, *The Suburban Gothic in American Popular Culture* (Basingstoke: Palgrave Macmillan, 2009), p. 200.
63 Beuka, p. 4.
64 Theodor W. Adorno, *Notes to Literature*, vol. 1, trans. by Shierry Weber Nicholsen (New York: Columbia University Press, 1991), p. 207.
65 It is worth noting that the particular suburb that Tavi grew up in, Oak Park, is not a typical American suburb. Home to architect Frank Lloyd Wright, it is not only architecturally distinct, but also largely liberal leaning and home to a lively art scene. However, since the setting of most of Tavi's photographs does not focus on these particularities but on a more generic suburban scene, it could be said that the environment that preoccupies her is American suburbia more generally.

66 Benjamin, 'Unpacking My Library', *Illuminations*, p. 61.
67 Tavi Gevinson, 'Editor's Letter – December 2013: Forever'.
68 Gevinson, 'Idea's At The House: Tavi's Big Big World'.
69 Benjamin, 'Unpacking My Library', *Illuminations*, p. 59.
70 Timotheus Vermeulen and Robin Van Den Akker, 'Notes on Metamodernism', *Journal of Aesthetics and Culture*, 2.0 (2010), www.aestheticsandculture.net/index.php/jac/article/view/5677 [accessed 2 Nov 2016]; For a basic history of the term in contemporary cultural theory also see Seth Abramson, 'Metamodernism: The Basics', *The Huffington Post*, 12 December 2014, www.huffingtonpost.com/seth-abramson/metamodernism-the-basics_b_5973184.html [accessed 14 July 2016]; Vermeulen and Van Den Akker also run a webzine in which multiple commentators discuss metamodernism in relation to a range of cultural and aesthetic phenomena, www.metamodernism.com [accessed 2 Nov 2016].
71 Nicholas Bourriard, *Altermodern: Tate Triennial 2009* (London: Tate Publishing, 2009); Kirby, Alan, *Digimodernism: How New Technologies Dismantle the Postmodern and Reconfigure Our Culture* (New York: Continuum, 2009); Svetlana Boym, 'Nostalgic Technology: Notes for an Off-Modern Manifesto', www.svetlanaboym.com/manifesto.htm [accessed 2 November 2016]; Lipovetsky, *Hypermodern Times.*
72 Vermeulen and Van Den Akker, 'Notes on Metamodernism', p. 5.
73 Timotheus Vermeulen and Robin Van Den Akker, 'What Meta Does and Does Not Mean', *Notes on Metamodernism*, 14 October 2010, www.metamodernism.com/2010/10/14/what-meta-means-and-does-not-mean/ [accessed 2 Nov 2016].
74 Vermeulen and Van Den Akker, 'Notes on Metamodernism', p. 2.
75 Vermeulen and Van Den Akker, 'Notes on Metamodernism', p. 5.
76 Abramson, 'Metamodernism: The basics'.
77 Wallace, David Foster, 'E Unibus Pluram: Television and U.S. Fiction', in *A Supposedly Fun Thing I'll Never Do Again: Essays and Arguments* (London: Abacus, 1998), pp. 21–82 (pp. 67–68).
78 Tavi Gevinson, 'Heart n Soul', *Style Rookie*, 17 November 2011, www.thestylerookie.com/search?updated-max=2011-12-12T16:29:00-06:00&max-results=1&start=22&by-date=false [accessed 4 July 2014].
79 Gevinson, 'Idea's At The House: Tavi's Big Big World'.
80 Gevinson, 'Idea's At The House: Tavi's Big Big World'.
81 Jameson, '*The Shining*', pp. 117–118.
82 Gevinson, 'Idea's At The House: Tavi's Big Big World'.
83 Gevinson, 'Idea's At The House: Tavi's Big Big World'.
84 Sontag, *Against Interpretation*, p. 312.
85 Ellen Willis, 'Three Elegies for Susan Sontag', in *The Essential Ellen Willis*, ed. by Nona Willis Aronowitz (Minneapolis: University of Minnesota Press, 2014), pp. 464–473 (p. 466).
86 Gevinson, 'Idea's At The House: Tavi's Big Big World'.
87 Wallace, p. 81.
88 Gevinson, 'Why The 21st Century Isn't All That Bad'.
89 Gevinson, 'Idea's At The House: Tavi's Big Big World' (my italics).
90 Gevinson, 'Strange Magic'.
91 Jean Baudrillard, *Simulacra and Simulation*, trans. by Sheila Faria Glaser (Ann Arbor: University of Michigan Press, 1994), p. 6.
92 Baudrillard, *Simulacra and Simulation*, pp. 2–3.
93 Higson, p. 126.
94 Gevinson, 'Five Years'.

Interlude 2

A space outside: utopia as negation

The moment of liberation which is betrayed by the brother clan – the memory of which is considered so powerful by Marcuse (see Chapter 3) – can alternatively be conceived as a moment of possible negation: of resistance, that is, to the totality of 'what currently is'.

Jameson makes explicit this link between the betrayed moment of liberation and what he sees as the impossibility of negation in a consumer, capitalist society. He argues that in late capitalism all potentially resistant energies or movements are incorporated into the mainstream: potentially subversive ideas, for example, are denigrated and robbed of their power by being turned into fads.[1] Mark Fisher agrees, insisting that 'alternative' and 'independent' no longer designate something outside of mainstream culture, but rather styles within the mainstream.[2] Fisher mentions as an example the fact that anti-capitalist narratives are widely disseminated in Hollywood (*Avatar* and *The Hunger Games*, both released after Fisher's book, are prime examples of this), and suggests that these films perform our anti-capitalism for us, allowing us to continue to consume with impunity.[3] Raymond Williams famously made a distinction between alternative and oppositional cultures. The former exist alongside the dominant culture but only the latter challenge the dominant culture.[4] To borrow Williams' terminology, Jameson and Fisher certainly argue that genuinely oppositional cultural practices are very rare or even impossible in late capitalism, but they also suggest that even alternative cultures are robbed of some of their heteronomous power by being somehow incorporated into the mainstream.

There are, today, subcultures or cultural communities that could be described as genuinely oppositional. ISIS, or IS or Daesh, comes to mind as a radically oppositional movement, with at least some, albeit minimal, local support in the USA; and teen shooters – whose ranks swell year on year – might be seen as a kind of oppositional subculture. But it is hard to conceive of either group as in any way utopian in a Blochian sense. Anonymous, the loose network of hackers, constitutes a cultural community that some might argue is both oppositional and utopian. However, the extremity of these examples suggests that there is merit in Jameson and Fisher's perspective that many potentially oppositional or even alternative cultures have been successfully incorporated into the mainstream. Tavi, for example, struggles with her sense that 'there used to be hippies and beatniks, now there's nothing!'[5] She also wonders if 'maybe there's no need to create a counterculture'. Interestingly, her conception of a counterculture as something that often helps 'to create just another standard of coolness, or something' demonstrates how far the idea that a counterculture could be more than 'just another standard of cool' has receded. There is very little sense that it could be driven by genuine or powerful resistance to contemporary capitalism.

This incorporation of potentially resistant energies defuses the very possibility of subversion represented for Marcuse by the forgotten moments of liberation that the 'brother clan' betrayed. What has been forgotten, in other words, is not just the possibility of successful liberation, but even the possibility of any kind of conceivable negation.

Jameson also argues that so-called tolerance in contemporary society is actually, in a subtle way, repressive, because the individual in this consumer society is 'denied that path towards genuine psychic individuality once offered him by the revolt against the father'.[6] If everything is on one level permissible, then revolt is no longer possible. Perhaps even more troublesome is the sense that although one does not have to behave or feel a certain way, one should want to. Žižek explores this issue and critiques the liberal-choice command: 'You may! Enjoy!' For Žižek, postmodern 'totalitarianism' rests on a supposed freedom which only deepens the sense of duty by demanding willingness. As he explains, it is now 'not only: You must visit your grandma, whatever you feel', but: 'You must visit your grandma, and you must be glad to do it!'[7] Where revolt or negation was possible in a society that overtly laid down the rules, in a society that claims not to do so the demand is often simply internalised. Another example of this is the contemporary command to 'have fun': the kind of intoxicated frivolity described by Benjamin in 'The Lives of Students' (see Interlude 1) is arguably more forcefully mandated in youth or student culture than any enforced learning. The following lines in Swedish singer Lykke Li's song 'Everybody But Me' testify to the frustrated, even impotent, desire for revolt:

> When everybody's dancing (I don't want to)
> When everybody's toying (I don't want to)
> When everybody's laughing (I don't want to)
> Everybody but me

The erotic realm is another area in which the possibility of negation seems to have receded further and further into the distance, the more surfacely tolerant society has become. Marcuse suggests already in the '50s that the relaxation of sexual taboos compared with the Puritan or Victorian eras has not in fact liberated the powerful force of a repressed Eros. On the contrary, so-called sexual liberty has only enhanced the infiltration by business interests of this previously hidden sphere: now, 'in their erotic relations, they "keep their appointments" – with charm, with romance, with their favourite commercials'.[8] Marcuse's commentary on the falsity of the so-called liberation of sexuality seems all the more prescient and compelling since the advent of the internet, with young teenage girls mimicking the stars of the omnipresent multi-billion dollar pornography industry in their body hair (or lack thereof) and 'sexting' poses. Of course, sexuality and business interests have been interwoven since the first act of prostitution, but Marcuse's point is that there were at least pockets of forbidden erotic encounters which were accordingly free, which the opening of sexuality into daily life (and thus commercial life) has all but eradicated:

> It could not be said of any of the sexy women in contemporary literature what Balzac says of the whore Esther: that hers was the tenderness which blossoms only in infinity.[9]

An additional aspect of the utopian to consider, then, is the possibility of negation. Just as the utopian resists instrumental reason (see Interlude 1), it can also be described as that which reaches towards negation, the conceptual possibility of a space outside of 'what is', in which what currently is is not mistakenly identified as what necessarily is and therefore always will be.

Notes

1 Jameson, *Marxism and Form*, p. 109.
2 Fisher, *Capitalist Realism*, p. 9.
3 Fisher, *Capitalist Realism*, p. 12.
4 Williams, pp. 10–11.
5 Casey, 'Rookie of the Year, Again: An Interview with Tavi Gevinson'.
6 Jameson, *Marxism and Form*, p. 109.
7 Slavoj Žižek, 'You May!', *London Review of Books*, 18 March 1999, www.lrb.co.uk/v21/n06/slavoj-zizek/you-may [accessed 15 May 2015].
8 Marcuse, *Eros and Civilization*, pp. 94–95.
9 Marcuse, *One-Dimensional Man*, p. 81.

5 Memories of longing in *The Virgin Suicides*

A vital part of the nostalgia boom first diagnosed by cultural theorists in the late 1970s was the 'nostalgia film'. It is telling that Fredric Jameson introduced his broad critique of the nostalgia 'mode' in then-contemporary culture via an analysis of the 'nostalgia film', which he saw as a central manifestation of this wider cultural phenomenon. His diagnosis of a culture hooked on the 'cult of the glossy image' understood the visual medium of film as in many ways ideal for propagating this new consumable nostalgia.[1] Christine Sprengler (2009), in her book on nostalgia in contemporary American film, observes:

> The Hollywood 'nostalgia film' is something that must be situated within the broader context of contemporary cultural nostalgia. One cannot come to grips with cinematic nostalgia without first understanding nostalgia's more general social significance and manifestations across visual (and political) culture.[2]

Inverting this, it is also the case that the visual, cultural and even political experience of contemporary nostalgia culture cannot be understood independently of the technologies – photography, television, film and the internet – that are its media.

Two waves of the American 'nostalgia film' are noteworthy. The first substantial wave of nostalgia films and TV sitcoms occurred in the 1970s and presented a nostalgic retrospective vision of a more innocent time, the 1950s, the period that Jameson has identified as 'the privileged lost object of desire'.[3] These included *The Last Picture Show* (1971), *Two-Lane Blacktop* (1971), *Grease* (which appeared first as a theatrical production in 1972, and was then released as a movie in 1978), *American Graffiti* (1973), *Happy Days* (1974–1984) – referred to by one contemporary commentator as 'basically *American Graffiti: The Sitcom*',[4] *The Lords of Flatbush* (1974), and *Laverne and Shirley* (1976–1983). The syndication of actual 1950s sitcoms such as *Father Knows Best* (1954–1960) and *Leave it to Beaver* (1957–1963) in the 1970s further contributed to the trend of filmic and televisual 1950s nostalgia.[5]

The second significant wave of nostalgia films appeared in the 1990s, when many films were made depicting 1970s life in a nostalgic way. Perhaps the most widely recognised 'nostalgia film' of this period is Richard Linklater's *Dazed and Confused* (1993) which is set in 1976 and, like *American Graffiti*, follows a group of teenagers on their last day of high school. *Dazed and Confused* has been pointedly referred to as a remake of *American Graffiti*, and although there are substantive differences in tone and intent, the two films are often considered together.[6] Other 1990s films that deserve consideration as 'nostalgia films' include *Summer of Sam* (1990), *Crooklyn* (1994), *Casino* (1995), *Boogie Nights* (1997), *54* (1998), and *Fear and Loathing in Las Vegas* (1998) (which all depict a gritty/glamorous urban 1970s) and *My Girl* (1991), *The Ice Storm* (1997) and *The Virgin Suicides* (1999) (which all, like *Dazed and Confused*, take 1970s suburbia as their setting).

The differences between these two waves of nostalgia films will be discussed in this chapter, but should not mask their many similarities, or the fact that nostalgia films and sitcoms have continued unabated both between the two waves and since the 1990s. In between the two waves, for example, the sitcom *The Wonder Years* (1988–1993), which depicts the 1960s, is undoubtedly a nostalgia sitcom, as *Dirty Dancing* (1987, set in 1963) and *Stand By Me* (1986, set in 1959) are nostalgia films. Recently, the most obvious example of a nostalgia programme is *Mad Men* (2007–2015), which follows characters from the late 1950s through to the early 1970s and is frequently cited in discussions about contemporary nostalgia culture, as by the following journalist:

> For all its retro appeal, *Mad Men* is essentially one long meditation on the central obsession of our time: nostalgia. With North American culture Periscoping and Snapchatting and Google-Glassing and Apple-Watching its way towards its post-human techno dystopia, nostalgia has become the opiate of choice for the digital masses.[7]

Other recent examples include *Wet Hot American Summer* (2001, set in 1981), *Adventureland* (2009, set in 1987) and the second season of *Fargo* (2015–2016, set in the 1970s) which, like *Mad Men* (although less overtly so), is arguably both an object of nostalgia for viewers and a critical meditation on the nature of nostalgia. These examples suffice to make the 'obvious point', as Sprengler argues, that 'nostalgia remains a vital part of contemporary life. The nostalgia thought to have been activated by fin-de-millennium anxieties certainly has not subsided'.[8]

Sofia Coppola's *The Virgin Suicides*, part of the second wave of nostalgia films, is a 1999 film. It is a close adaptation of Jeffrey Eugenides' 1993 novel of the same title. The story is set in Grosse Pointe, a Michigan suburb, in the 1970s. It concerns five sisters, the Lisbon girls, who commit suicide (we are told this, the story's end, in the opening scene of the film). Perhaps more accurately, though, it concerns the fascination and allure that these sisters exerted for a group of neighbourhood boys. The film is narrated by one of the boys (we never find out which), now a man, looking back on the period before and after the Lisbon girls' suicides. It explores the boys' experience of having longed to know the Lisbon sisters during adolescence – dreaming of, and desiring, them – as well as their subsequent desperate attempts to understand why these young girls had taken their own lives. The narrator tells us that the boys, now men, remain haunted by the memory of these enigmatic sisters and the mystery of their deaths.

There are many reasons for considering *The Virgin Suicides* – both book and film – as nostalgia texts.[9] Jeffrey Eugenides grew up in Grosse Pointe, the suburb outside of Detroit in which *The Virgin Suicides* is set, during the period in which it is set (the 1970s), and the narrators in the novel are of a similar age to Eugenides, and look back on a teenage experience that is presumably close to his own. The novel offers a complex meditation on longing, memory, fantasy and nostalgia, and is pervaded by an elegiac tone, which has been loyally recreated in the film. Just as most reviewers and critics of Eugenides' novel call attention to either the 'elegiac' or 'nostalgic' quality of the prose, reviewers of Coppola's film rarely fail to mention the nostalgic quality of the story and cinematography. For example, a 1993 review of the novel remarks on 'a tone that is both elegiac and hilarious' and mentions 'Eugenides' assured mix of heartfelt nostalgia – suburban life has seldom been recalled so lovingly – and dark humour'.[10] Dines describes the novel's 'insistent elegiac tone' and the 'very palpable nostalgia that pervades the novel'.[11] Similarly, a 2001 review of Coppola's film discusses its 'nostalgic glam'[12]; a 2012 review refers to it as a 'cult classic' and mentions its 'eerie nostalgia'[13]; and a 2015 retrospective discussion of the film makes clear that 'this is a nostalgia piece, about boys looking back on the fever-dream of adolescence and the fantasy of the girls next door, rather than a hard-edged tragedy observed in the moment.'[14]

The nostalgia of *The Virgin Suicides* functions on several levels. First, it is a nostalgic depiction of adolescence. Second, it is a nostalgic depiction of 1970s suburban America, of a time or kind of society that is being (or has been) lost. Heusser (2007), writing about the novel, argues:

> At one fell swoop the strangely disembodied collective narrators lose not only their innocence – their own youth and childlike innocence – but they also lose an innocent America, to which they hark back with intense nostalgia: a more 'real' America, one with actual winters, 'vast snowdrifts, days of canceled school.'[15]

I will argue that it is something more complex than innocence which is longed for, or lost. Nevertheless, Heusser's claim that the nostalgia evident in the book is both for a particular time of life and for a different kind of society is apt, and equally true of the nostalgia that permeates the film. Scenes typical of adolescent experience (going to prom, holding hands in a cinema, doodling in a diary, making out in a car) are lent a particular frisson by their purposively retrospective setting in a moment – before the arrival of the internet and still firmly in the grip of in the Cold War – when wallpaper was garish and families still gathered around 'the box' for their evening entertainment.

Third, at the time of their release both book and film seemed to evoke feelings of nostalgia in readers and audiences. For example, one contemporary reviewer writes that 'at the time I watched the film, a great, almost unsayable yearning was moving through my blood, nostalgia for moments that never existed'.[16] The novel, too, belonged to the particular constellation of the 1990s nostalgia wave, as is suggested by the following blogger (in a post about both film and book):

> In late 1999 I was sixteen years old, still developing an appreciation for films and books and music [...]. Here was an author with an unusual but striking voice who was able to pitch directly into teenage ennui, as well as conjure nostalgia for a time I never knew. Reading that book, listening to Air and Pink Floyd, and having already swooned to Richard Linklater's *Dazed and Confused*, I was a teenager in the 90s wishing he'd been a teenager in the 70s.[17]

The final reason for analysing *The Virgin Suicides* as a nostalgia text, the film and, consequently, the book also, have since their initial releases also enjoyed a second wave of nostalgic reception in recent years, particularly among teenage girls. Tavi Gevinson has many times written about her fascination with *The Virgin Suicides*, and often posted images from the movie on her blog. She writes that '*The Virgin Suicides* as a unit – book and movie – has influenced the way I think more than most things in life'.[18] Tavi's enthusiasm for the book and movie struck a wider chord, and she may well have affected the second wave of enthusiasm for *The Virgin Suicides* (*Time* magazine named her as one of the twenty-five Most Influential Teens in 2014).[19] Her influence is discussed in the following excerpt from an interview with director Sofia Coppola, conducted by Tavi for *Rookiemag*:

TAVI: Our theme on the site this month is Longing, and I think your films capture that feeling so well, and I have to ask a *Virgin Suicides* question because –

SOFIA COPPOLA: OK! I love that you brought that movie to a whole new audience! When I met Leslie Mann for this movie she said, 'Oh, my daughter loves *The Virgin Suicides*,' and I thought, *How does she even know about that? She wasn't even born then!* And it's through you, so –

TAVI: Ah!

SOFIA COPPOLA: That movie didn't have a big audience when it came out, so I'm happy that it lived on.

TAVI: Oh my gosh – girls WORSHIP it. It's like [...] it comes up so much when we meet our readers.[20]

Both at the time of release, then, and in its second wave of reception nearly fifteen years later, *The Virgin Suicides* elicited nostalgia in its viewers: particularly, it seems, and contrary to what one might expect, among those who are or were teenagers at the time of viewing. Initial consideration of *The Virgin Suicides* as a nostalgia film is thus warranted both on the primary grounds of its setting and narrative, but also with regard to its reception.

Jameson and the nostalgia film

As discussed, the 'nostalgia film' was for Jameson a central manifestation of the wider 'nostalgia mode' that was introduced and discussed in Chapter 1. As Sprengler points out, 'a quick glance at the scholarship confirms that Jameson's discussion of *American Graffiti* is the seminal text on the nostalgia film'.[21] Jameson's central criticism of nostalgia films, mirroring his criticism of the nostalgia mode in general, is that they efface and commodify history: rather than presenting a coherent or realistically complex representation of a given historical moment, nostalgia films present the past as nothing other than a series of aesthetic styles made available for easy consumption in the present. Past eras are represented by superficial visual signifiers; for example the 1950s becomes the 'Fifties'. According to Jameson, the emergence of the 'nostalgia film' in the 1970s, as a staple of the Hollywood diet, must be understood in relation to a broader crisis in historicity and a lack of depth in postmodern culture (see Chapter 1). Nostalgia films peddle an intentionally emotive trip down a cultural memory lane and should be understood as vehicles for a consumable nostalgia that nullifies any serious or critical engagement with memory or the past.

In addition to this attitude to history (his primary critique of the 'nostalgia mode') Jameson explores several other defining features of nostalgia films. One of these is their relationship to pastiche. Jameson distinguishes pastiche from parody and argues that pastiche, like parody, mimics previous works, but where parody did so to make a point of its own, pastiche does so to no discernible end:

> Pastiche is, like parody, the imitation of a peculiar mask, speech in a dead language: but it is a neutral practice of such mimicry, without any of parody's ulterior motives, amputated of the satiric impulse, devoid of any laughter and of any conviction that alongside the abnormal tongue you have momentarily borrowed, some healthy linguistic normality still exists.[22]

In a potted history of film, Jameson lays out several stages of filmic history: first, *the masterpiece*, early attempts at distinctive individual creation guided by a single personality; second, *genre films*, in which he includes musicals, westerns, *film noir*, and the classic Hollywood comedy or farce; third, *metageneric production*, in which he includes the war movie, the occult movie, the thriller, the western, science fiction, the musical, and the spy movie. Although Jameson does not give any clear criteria for distinguishing between *genre films* and *metageneric production*, and although the western and the musical appear under both headings, it is clear that he is more damning of metageneric production.

The supposition is that genre films had something of their own to say and that this is no longer the case with metageneric productions, the primary aim of which is financial success. Metageneric films draw on existing genres only insofar as they appear to exhibit an already successful formula, not to their own ends. As evidence of this he cites his own formative experience of film:

> No one whose life and imagination were marked and burned by the great images of *film noir* or infected by the immemorial gestures of the western can doubt the truth of this for a moment.[23]

Jameson here sounds disconcertingly ungenerous about contemporary film-making: several of the films he decries as mere metageneric production have gone on to become genre classics in their own right, engendering as much loyalty in subsequent generations of film buffs as the films of his youth did in his, for example *Chinatown* and *Dr Strangelove*. However, his distinction between films that are driven to tell their own truth and those which simply mimic existing genres is worth considering, regardless of whether his application of this distinction in different cases is accurate. Quentin Tarantino's films might serve as a good example of what Jameson means by 'genre pastiche' or metageneric production (although Jameson doesn't himself discuss Tarantino): they refer endlessly both to older films and to each other, exhibit high levels of what Mark Fisher calls 'PoMo swagger', and do not appear to have any interest in, or even remembrance of the possibility of, the modern project of unearthing truth or exposing fallacy.[24] For Jameson, the 'nostalgia film' is a 'degraded version' of metageneric production. Nostalgia films, like any metageneric film, often trade on pastiche and intertextuality, but they go one step further and attempt pastiche not only of other films or genres but of history itself.

Another defining feature of many nostalgia films, for Jameson, is beauty. He says: 'What is inauthentic about nostalgia films [...] can best be dramatised in another way by what I will call the cult of the glossy image.'[25] It is not that Jameson objects to beauty per se, for example he cites Van Gogh's flowers as an example of how intense beauty has often been 'a wrenching political act'.[26] He objects, rather, to the application of technical mastery to create images of overwhelming beauty that express nothing other than mere surface. In response to filmmaker Terence Malick, for example, he says:

> Is it ungrateful to long from time to time for something both more ugly and less proficient or expert, more home-made and awkward, than those breath-taking expanses of sunlit leaf-tracery, those big-screen flower bowls of an unimaginably intense delicacy of hue, that would have caused the impressionists to shut their paint boxes in frustration? I hope it is not moralizing to admit that from time to time such sheer beauty can seem obscene, like some ultimate in the consumption of streamlined commodities – a transformation of our senses into the mail-order houses

> of the spirit, some ultimate packaging of Nature in cellophane of a type that an elegant shop might wish to carry in its window.[27]

In criticising the 'culture of image', it might be said that Jameson has two things in mind. He means, first of all, a culture that propagates photographic and filmic images at a high intensity; and, second, a culture of superficiality: one in which the 'image' of a thing supersedes, or indeed erases, its truth.

What Jameson calls the 'cult of the glossy image', and for which he blames the lavish indulgence made possible by 'a whole new technology (wide-angle lens, light-sensitive film)', is even more widespread today thanks to the development of digital photography. One thinks, for example, of the many Instagram nature photographers who post daily 'inspiration', with hashtags such as #shotsofawe: images that might once have actually awed viewers (and taken careful months for a photographer to capture) now arrive unbidden into one's iPhone, a stream of ubiquitous mastery about which the phrase 'mail-order houses of the spirit' seems ever-more pertinent, excepting only the now somewhat quaint anachronism of a *mail* order.

A Jamesonian reading

The opening scene of Coppola's *The Virgin Suicides* is the archetypal suburban scene, so familiar to viewers of American film and television that it conjures up at once a complex web of meaning and association. The film begins with the camera panning over a neighbourhood which, by the use of a number of familiar markers that Robert Beuka has called 'loaded signifiers', presents itself as exactly the suburbia audiences know to expect: wide roads, the sounds of birds and crickets suggestive of summer heat, sprinklers at work on the carefully manicured lawns, dogs barking, kids playing basketball in the driveway of their house, the sun shimmering through dappled green leaves, and a pretty Lolita-esque teenage girl sucking a popsicle.[28] Beuka observes that from early on in the massive post-war development of suburban housing in the USA, TV helped to invest the emergent landscape with symbolic meaning: the expansion of suburbia entailed 'the construction of not only a new physical landscape, but new psychic and emotional landscapes as well'.[29] It is precisely this psychic landscape – rather than any geographical or historical particularities – that Coppola invokes. David Coon argues that nostalgia is often used in the promotion and development of this psychic suburban landscape, and describes:

> the complex process whereby nostalgia is used to create suburban imagery, which is in turn used to create a sense of nostalgia, binding the two together in a simulacral relationship that suggests that American suburbia exists as little more than a set of intertwined images, incessantly replayed for our consumption.[30]

Given this, the suburban imagery that Coppola presents us primes the viewer straightaway for a kind of nostalgic reception – as much for previously consumed ideas and images of 'suburbia' as for any lived reality.

Appearing over the images of this suburban scene, the opening titles spell out the pastness of the movie – 'Michigan. 25 years ago' – further inviting the viewer to enter into a nostalgic mode of reception. This scene-setting title is scrawled in bubbly teenage handwriting, replete with hearts and doodles as if in a girl's diary or notebook, giving an early indicator not only that this is a film about teenage girls but that it is to be enjoyed. Had the film been intended as tragic (which, given its core subject – the suicides of five sisters – it might have been) it would not have opened with cutesy teenage handwriting and intermittent images of a girl floating on a cloud and winking.

Moreover, although we are told the narrative's rough location in place (Michigan) and time (twenty-five years ago), the film is intentionally blurry with regard to specifics. Viewers of the film when it was first released in 1999 (2000 in some places) would have been able to date it to 1974 or 1975, but Coppola will have been aware that on later viewings the release date of the movie would be unlikely to come to mind. Instead, two important ideas are communicated with this opening scene-setting: this is a movie about remembering a time gone by ('25 years ago' achieves a similar effect to the 'a long time ago' or 'once upon a time' of children's stories) and it is set sometime in the mid-1970s – or rather, in Jameson's language, the mid-Seventies. The set and costume design clearly evoke the Seventies – a brown-ey pink-ey colour palette, garish yet weathered flowery wallpaper, deep V collars and flares – but the film makes little reference to any cultural or political particulars. Film critic Keith Phipps also mentions this historical and geographical vagueness:

> The production design and period detail is remarkable while also being kind of non-specific [...] it's in the mid-1970s, but it doesn't feel tied down to any specific moment. Similarly, it's firmly in the affluent (but declining) suburbs of Detroit, but it also feels like it could be a suburban, Midwestern anywhere.[31]

It is clear, then, that although the film is set in the past Coppola does not create a faithful history of a moment: the concern of the film is rather the past of memory, the past from a fuzzy dream. Coppola explains her intention in an interview: 'Jeffrey calls the Lisbons the fever dream of the boys. I wanted to make the movie a fever dream.'[32] Insofar as this is a story about grown men remembering their youth, it is certainly legitimate to give the film the air of a 'fever dream', since this captures something about the nature of memory. However, the film does more than this. It brings viewers into the perspective of the 'rememberer', asking them to come along and enjoy the nostalgic ride.

One of the means by which the film invites viewers to feel nostalgic is the use of voice-over narration. The film is narrated throughout in the voice of a grown man remembering his youth. It is unclear which of the neighbourhood boys the grown-up narrator is meant to be: he speaks in the first person plural ('we') to articulate their collective memories of the summer when the Lisbon girls committed suicide. One of the innovations of Eugenides' novel was to narrate it in the first person plural and, although something of the anonymity and collectivity of this literary technique is perhaps lost in Coppola's film adaptation, since we hear only a single narrator's voice, Coppola uses other techniques to anonymise the narration. The audience is given no information about the narrator in his adult life. In the 1986 film *Stand By Me*, which also looks back nostalgically to the experiences of four boys during a summer several decades past, we see the narrator (now a middle aged man, a writer) both at the start and the end of the film. We also know which of the boys he is, thus giving context to his particular perspective and memory. In *The Virgin Suicides*, by contrast, we never see the narrator, nor learn anything about the particularities of his current situation. Furthermore, the voice – a soft, husky voice – is of unclear age. Performed by the actor Giovanni Ribisi, who would have been twenty-three or twenty-four when filming took place, the narrator's voice does not seem to be that of a teenager or of a middle-aged man. Instead, it seems caught at an age somewhere between that of the boys he is remembering, and that of the men they have become. With the particulars of the narrator sufficiently blurred, the voice thus functions as a kind of generic voice of remembrance. If in *Stand By Me* the audience is able to identify the narrator and locate his process of memory, thus clearly distinguishing their own position from the man whose memories they are watching, in *The Virgin Suicides* the voice-over frames the entire experience of the film: with no distancing devices, the perspectives of audience and narrator begin to merge.

The cinematography is also striking in its targeted production of nostalgia, further suggesting that *The Virgin Suicides* is intended to be consumed as a nostalgia piece. Several critics and reviewers comment on how the director of photography, Edward Lachman, deliberately creates a nostalgic mood that emotionally impacts the viewer. Hoskin, for example, mentions how Lachman 'shoots the bright colours of the 1970s as if through a layer of gauze'.[33] More pointedly, at least two reviewers liken the hazy imagery that Lachman achieves to that of old photographs. One reviewer argues:

> Edward Lachman's gorgeous cinematography contributes to the film's nostalgic glam [...] scenes from the film (first kisses, gossiping about neighbours) are sinewy in nature and seem lifted from the pages of a lost photo album.[34]

Another suggests that 'the whole film has a blurred and fuzzy look to it, like looking at an old photograph' and points out that 'the sound design plays into this sense of nostalgia as well, we can often hear the dead space on a record for no discernible reason at all.'[35] That the film intends to evoke nostalgia is not lost on viewers. As critic Peter Preston observes of his own experience watching *The Virgin Suicides*:

> As you sit there, your own memories come flooding back. I remembered the girl called Jaqueline of long ago who died, trapped in her bedroom, when the house caught fire, and I remembered how I felt. I was meant to remember, invited to look back.[36]

A further reason for reading *The Virgin Suicides* as just the kind of nostalgia film that Jameson decries is that in look and feel the film in many ways resembles an extended music video or mood piece. This is most obviously the case with the many extended scenes that break the flow of the film in order to present what are perhaps best described as musically accompanied visual nostalgia shows. Colin Tait, in an article about the second wave of nostalgia films (1990s films that were nostalgic for the 1970s), points out how often they include what he calls the '1970s sequence', scenes which 'absorb the logic of the music video'.[37] He describes these scenes as follows:

> More often than not, these sequences disrupt the coherence of the narratives, presenting what is essentially a music video to a 70s song. While loosely inspired by the events within the movie, they often move outside of their narrative frameworks to express a totally different logic, favouring style over formal structure.[38]

He then goes into further detail about the specific techniques typifying these scenes:

> Rapid-fire cutting, temporal manipulation (usually slow-motion), and over-processed and tinted film stocks together comprise a remarkably standard stylistic of the present's nostalgic gaze upon the 70s. All of these techniques fetishize a lost era's cinematic style while paradoxically evoking a contemporary visuality supplied by modern (often digital) technology.[39]

Tait mentions several examples of this (the 'tiny dancer' reconciliation in *Almost Famous* (2000), the opening credit sequence in *Reservoir Dogs* (1992), the disco/porn montages in *Boogie Nights* (1997), the montage of 70s iconography choreographed to The Who's 'Baba O'Riley/Teenage Wasteland' in *Summer of Sam* (1999)), including the scene of the adolescents playing records in *The Virgin Suicides*. In this, the girls are locked up at home by their strict mother, and the boys call them on the phone and play records to them. Different 70s tracks blend into one another in this extended scene which includes split screen views of the boys and the girls – apart yet together – and close-ups on the faces of each. Tait's insight that this scene both resembles a music video and is, like others of its time, an almost textbook example of cinematic pastiche, is important. The Jameson-esque flavour of his analysis is unmistakable:

> The curious fascination with this raw and unprepossessing recent past, compounded by an aesthetic that betrays its own present-tense*ness*, endows these films with a schizophrenic sense of waning affect.[40]

There are several other similar scenes in *The Virgin Suicides*. For example, the moment in which the girls lounge in their bedroom reading exotic magazines and dreaming of faraway places, the camera lingering on the chintzy details of their bedroom, filled with panties, lipsticks and cheap Christian iconography; or the extended slow-motion tracking of Trip Fontaine, the attractive if obvious boyish rebel, lolling down the school corridor; or another where the camera lingers lustfully from a bird's-eye view on the svelte and bronzed Trip lying on a lilo, wearing tight patterned swim trunks and sipping a pink cocktail, comically reminiscent of a Seventies porn star. The frequency of these kinds of scenes means that, taken as a whole, the movie is almost a visual scrapbook, patching together 'loaded signifiers' of both the Seventies and of suburbia and offering them up to an audience it has primed for nostalgic reception.

Tait also points out that 'these scenes are often more memorable than the movies themselves'.[41] This is an apt description of how *The Virgin Suicides* is remembered and consumed among its most recent wave of teenage girl enthusiasts. Short clips or film stills of the aforementioned scenes (and others which follow a similar logic) circulate widely, endlessly replayed both on Tavi's blog and on countless other teen blogs and tumblrs. For example, http://virgin-suicides.tumblr.com/ is devoted exclusively to film stills, short video clips, songs and quotations from the film; and indeed perusing the tumblr doesn't feel too dissimilar an experience to watching Coppola's film. Another tumblr, http://fuckyeahthevirginsuicides.tumblr.com/ is much the same.

Coppola's simultaneous work both as a fashion photographer and a film director is also worth considering, as is the mutual influence between the fashion world and *The Virgin Suicides* film. Regarding the influence of a fashion logic on the film, Coppola's background in fashion prior to becoming a film director can be said to have influenced the style of *The Virgin Suicides*, her first feature film. Mayshark elaborates this point:

> She specializes in atmosphere, conveying ideas through a combination of beguiling images, gliding camerawork and artfully deployed music. Her films feel designed as much as directed, which is not intended as a slur. She has a fashion photographer's eye for composition and suggestion (not a surprise, given her background as both an occasional model and a clothing designer), and a striking sense of the use of light.[42]

Tait points out that the 1990s saw the rise of a new generation of filmmakers who began their career in the advertising and music industries (David Fincher, Michael Gondry, and Spike Jonze – Coppola's Husband from 1999 to 2003).[43] Mayshark references a 'cultural tide of pop postmodernism that reached a peak in the music, film, literature and fashion of the 1990s'.[44] The relevance of this analysis is that Coppola was very much a part of her generation, always a fashion icon as well as a film director and enmeshed in a culture well-versed in postmodernism, and in which film and fashion were different spokes on a wheel, both considered a complex combination of art and commerce. One is reminded of the fact that Jameson saw fit to emphasise Andy Warhol's early career as a commercial illustrator for shoe magazines in his discussion of Warhol as a postmodern artist. Indeed, the fading distinction between high and low art, and more broadly between art and commerce, was a key factor in Jameson's account of postmodern culture.

Looking from the other direction (the influence of the film on subsequent fashion) is also informative. A 2015 article in *Dazed and Confused* magazine, titled 'Deconstructing the Fashion of *The Virgin Suicides*' describes how '*The Virgin Suicides* aesthetic has influenced an entire generation of bored teenage girls' and, more specifically, details its influence on fashion, not only via blogger Tavi Gevinson, but also elsewhere:

> The influence on the fashion world has been seen in the whimsical style of Rodarte, as well as Coppola collaborator Marc Jacobs (the *Daisy* campaigns, as directed by Coppola, are clearly a part of the Lisbon sisters' world).[45]

A comparison of stills from *The Virgin Suicides* film with the *Daisy* fashion campaign demonstrates this clear influence.[46] The strong similarities between the two are striking: dreamy imagery of girls lounging around in a field or on bed, wearing white to evoke a glow, soft hues suggestive of lazy summer days and, as often in Coppola's visuals, evocative of a heady mix of ennui and longing. Coppola directed the advertising campaigns, so some similarity in aesthetic might be expected, but the ease with which the scenes from the film are effectively transposed into the campaign suggests something about the way that they were consumed originally: as pages in a filmic scrapbook as much as moments that make sense only within the narrative drama of the film.

There is ample evidence, then, for reading *The Virgin Suicides* as a 'nostalgia film' that Jameson would have decried, had he analysed it. It accesses the past by way of 'stylistic connotation' rather than historical content, and its lasting influence has been in fashion and style. As such, we might well believe that it was motivated from the outset by the 'cult of the glossy image', and might even therefore regard it as a kind of extended music video or mood piece, a patchwork of Seventies suburban pastiche, peddling an enjoyable, if bitter-sweet, nostalgic rumination on adolescence, suburbia, and the long hot summers of an earlier, seemingly more innocent America.

Jameson versus Bloch: a methodological reflection

Thus far, I have characterised Jameson as exclusively critical of the 'nostalgia mode'. My readings, in this chapter and Chapter 4, begin by identifying 'the rubble', the ways in which these nostalgia texts are indeed products of depthless late capitalism in just the way Jameson describes. Since Jameson's core critique of the 'nostalgia mode' has had remarkable staying power, with critics today still regularly drawing on and grappling with Jameson's analysis, it seemed necessary to focus considerable energy here, both in showing how relevant that critique still is, and in looking to move beyond it.

However, Jameson's writing on the subject are, not surprisingly, more nuanced than his core position would suggest. Significantly, he does allow that nostalgia is sometimes used in more critically valuable ways. For example, he gives an extended reading of Stanley Kubrick's *The Shining* as a film that enacts elements of a 'nostalgia film' in order to make a critical comment on nostalgia films. He first describes the Jack Nicholson character (pointedly referred to as such by Jameson) as an 'atomized individual' existing in the banal world of the modern-day hotel, who is haunted by the previous inhabitants of its once grand, lively past. This haunting, he argues, demonstrates a kind of nostalgia for community, or collectivity:

> [...] this is, finally, what I think *The Shining* is all about. Where to search for this 'knowable community' [...]? It is surely not to be found in the managerial bureaucracy of the hotel itself, as multinational and standardized as a bedroom community or a motel chain; nor can it any longer take seriously the departing vacationers of the current holiday season, on their way home to their own privatized dwelling places. It has only one direction to go, into the past.[47]

However, if one looks more closely, Jameson goes on to point out, the ghosts that haunt the hotel are all from the 1920s – the last period in which class structure was proudly out in the open, in which 'the American ruling class led an aggressive and ostentatious public existence [...] and enjoyed itself without guilt, openly and armed with its emblems of top-hat and champagne glass'.[48] Jameson thereby argues that the film enacts what might at first seem to be an understandable nostalgia for community or belonging in an increasingly atomised world, but by focusing on the 1920s as the lost period *The Shining* exposes what lies at the heart of this nostalgia: a longing for the certainty of a more-rigid class structure. In other words, the film enacts the nostalgia mode only to expose the fact that nostalgia films are not merely harking back to a simpler, more innocent, or community-oriented time, but have more sinister ideological content. *The Shining* works by:

> implacably demystifying the nostalgia film as such, the pastiche, and reveals the latter's social content: the glossy simulacrum of this or that past is here unmasked as possession, as the ideological project to return to the hard certainties of a more rigid and visible class structure.[49]

Jameson's reading of *The Shining* does not negate his standing critique of most nostalgia films. Just as it is important to understand that his view of postmodernism does not exclude the possibility of critical resistance, so his diagnosis and critique of the nostalgia mode, which he sees as most often without merit and a carrier of dangerous ideological content, does not exclude the possibility that filmic or literary nostalgia might also be used, sometimes, to different effect. As he says in an interview about his analysis of postmodernism:

> It is important to understand that this notion of a dominant does not exclude forms of resistance. In fact, the whole point for me in undertaking this analysis was the idea that one wouldn't be able to measure the effectiveness of resistance unless one knew what the dominant forces were.[50]

Others have followed a similar approach to that in Jameson's analysis of *The Shining*. David R. Coon, for example, gives a reading of two suburban films, *The Truman Show* (1998) and *Pleasantville* (1998), in which he argues that both films appropriate the nostalgia mode in order to highlight its falsity: 'the past that people feel so nostalgic about is shown to be a fabrication'.[51] One of the means by which these films do this, he suggests, is by drawing attention to the unreliability of memory, thereby showing that it often leads to a false understanding of the world. The final message of these films, according to Coon, is that 'the present might be complicated but it beats the repression and repetition of a life stuck in the nostalgic past'.[52]

Vera Dika uses a similar perspective to question Jameson's analysis of *American Graffiti*. It is important to note that she does not disagree with Jameson's critique of the nostalgia mode per se; her attempt to salvage *American Graffiti* relies on an argument that the film was more conscious and critical in its enactment of nostalgia than Jameson allows (the same approach that Jameson himself takes to *The Shining*). Dika recognises the merits of Jameson's analysis of *American Graffiti*: 'Not only does *American Graffiti* return to the styles and songs of a trouble-free pre-1960s era, one before the political disruptions of the 1960s proper, but it also embodies an apparently non-critical, pleasurable style of film-making.'[53] But she argues that the film, made in 1973 and set in 1962, relies on its audience's remembrance of the subsequent historical events (President Kennedy's death in 1963, the Vietnam War, the civil rights movement and political upheaval of the later 1960s) to ensure that the 'innocence' presented in the film is experienced with a certain foreboding:

> The '1960s' are thus fashioned to a particular meaning in *American Graffiti*. They are invoked to reference the effects of the Vietnam War on a generation, with the enforced gaiety of the film set against this historical loss.[54]

She reads the film therefore as not only nostalgic, but also, and at a more fundamental level, as shot through with pathos. Although the subsequent years are not referenced at all throughout much of the film, she suggests that they are nonetheless ever-present, inseparable from the then contemporary viewers' experience of the film. Only at the end of *American Graffiti*, when the closing titles describe what happened to each male character (their destinies each heavily influenced by the Vietnam War), is this unspoken (but assumed) knowledge made explicit. Showing some hesitation in her own analysis, Dika does allow for the possibility that *American Graffiti* does not do enough to ensure that the pathos she describes is registered by the viewer, and wonders if 'the easy pleasure of watching *American Graffiti* lulled the viewer into its illusionistic play without the critical distance needed for "resistance" in its true form'.[55]

In an interview about making *American Graffiti*, George Lucas spoke about having written the film at a time when Hollywood was making a number of dark, depressing films and explained that he had felt it was time for 'a movie where people felt better coming out of the theatre than when they went in'.[56] This seems to lend credence to a Jamesonian reading of the film as a vehicle for consumable, enjoyable, nostalgia, rather than, as Dika hopefully suggests, a critical comment on the pseudo innocence of an era in light of the disruptive events that followed. Dika's suggestion, for example, that the neckings and sexual fumblings represented in *American Graffiti* would not seem quaint to an audience familiar with that era's lack of effective birth control and sex education seems to assume an audience ready to politicise what is not presented in the film in any overtly political way, and is less convincing than the more straightforward assumption that, as Lucas intended, the audience felt good about the film, and revelled in nostalgia for the frisson of early sexual exploration: the forbidden yet relatively innocent act of teenage necking in the backseat of a car. Indeed, a happy nostalgic reception is described by one blogger, a teenager at the time of *American Graffiti*'s release, who recalls 'the good time cultural frenzy that the movie touched off [...] for a while, everyone played '50s music, danced '50s dances, and tried to pretend they were living inside *American Graffiti*'.[57] In the end, then, Dika's analysis does not sufficiently upset Jameson's reading of *American Graffiti* as a pure (or almost pure) example of the nostalgia mode.

But what is important from my perspective is that Jameson's analysis of *The Shining*, Coon's analysis of *The Truman Show* and *Pleasantville*, and Dika's analysis of *American Graffiti*, all attempt to rescue certain films that appear at first glance to be nostalgia films by arguing that they are actually only employing the nostalgia mode in order to subvert it. One can debate both the intention and success of the films in question, but all three critics are trying to claim something similar for the respective films that they analyse. What I am calling a Blochian method is quite different. In the Blochian reading I give I will not attempt to show that the nostalgia evident in *The Virgin Suicides* is a clever narrative device or mode of representation used to expose the problematics of historical representation or of the nostalgia mode itself. As discussed in Chapter 3, there is an important distinction between a critical method of exposition (in which the goal is to expose ideology) and a critical method of hermeneutics (which traces hidden moments or aspects of hope within ideology). Using the latter, in this instance, involves accepting as a starting point that *The Virgin Suicides* exhibits exactly the kind of ahistorical, unspecific, even sentimental enjoyment of pastness that Jameson is wary of – the kind which, as explored, continues easily in adverts and on blogs, evoking similar affective responses in those other media to those evoked in the film. However, by looking more closely at what exactly the nostalgia is for, I hope to show that there are traces of hope, of forgotten ghosts and dreams that need re-awakening, in the nostalgia itself. I will argue that, rather than merely evidencing a neurotic libidinal historicism in a world with no access to history, the nostalgia contains within it longings that are at odds with the present-day, and as such it highlights a genuine lack or, put differently, a hole within the whole, of present-day ideology: a dormant promise that might yet re-awaken.

A Blochian reading: the gold in the rubble

Several critics of Eugenides' *The Virgin Suicides* have argued that the grown men narrating the story are stuck in the past. Martin Dines argues that the men 'have failed to come to terms with the deaths; manifestly, they are still in mourning'.[58] Debra Shostak, similarly, suggests that the suicides of the girls 'stopped time for the narrators, who seem to live in a timeless zone of contemplation of the Lisbon deaths'.[59] There is some evidence to support this point of view in both the novel and Coppola's film. In the film, the voice-over narrator, although giving away next to nothing about the men's current lives, does, within the first few minutes of the film, describe the Lisbon sisters as that which 'after all these years, we can't get out of our minds'.[60] Later, in a line lifted directly from the novel, the narrator describes how memories of the girls 'have scarred us forever, making us happier with dreams than with wives'.[61] A passage from the novel which did not make it into the film but which helps flesh out the idea that these grown men are 'happier with dreams than with wives' is one in which the men admit that their memories and fantasies about Lux (the most sexually active of the sisters, whom the boys repeatedly see in trysts with different neighbourhood men on the roof of her house) over-shadow their real intimacies:

> [...] even now, if we were to be honest with ourselves, we would have to admit that it is always that pale wraith we make love to, always her feet snagged in the gutter, always her single blooming hand steadying itself against the chimney, no matter what our present lovers' feet and hands are doing. And we'd have to admit, too, that in our most intimate moments, alone at night with our beating hearts, asking God to save us, what comes most often is Lux, succubus of those binocular nights.[62]

It is easy enough to cast this dwelling in the past as regressive, a pointless longing that prevents the narrator(s) from moving forward (paralleling Coon's somewhat trite understanding of the nostalgia in *Pleasantville*, that it is used to show the audience that 'the present might be complicated but it beats the repression and repetition of a life stuck in the nostalgic past').[63] But what if we take a Blochian view that the past is sometimes the container of *still latent*, unfinished dreams? Before dismissing the narrator's nostalgia as limiting and regressive, something which should be revealed as a pointless longing, is it not worth looking first, in depth, at what, specifically, he is (or they are, since the voice speaks for the collective) nostalgic for? People dwell in the past for a number of reasons: sometimes, to be sure, it is simply an inability to move on, but is it not sometimes also a refusal to give up on something important? If the adult men are still dreaming of their earlier fantasies, what is it that they are seeking there? What promises have been disappointed in their adult lives, which still shimmer in their memories, 'insisting' as 'betrayed radical-emancipatory potentials', 'ghosts which need redeeming'?[64] What is it, alone at night, dreaming of Lux, that they are asking God to save them from?

The association between nostalgia and childhood is long-standing. Over the course of the nineteenth century the modern understanding of nostalgia as a longing for lost time emerged, and childhood was increasingly idealised as a distinct and special time of life. Jenß points out that in this period 'childhood began to be conceived as a time of innocence, purity and discovery' and explains that this was a reaction to rapid industrialisation and the changes that it wrought: 'Childhood became an important compensatory component in the modern conception of time and the experience of change and acceleration.' In other words, childhood 'was made over into the antithesis of the machine'.[65] As discussed in Chapter 1, modern nostalgia also arose partly in resistance to increasing speed and industrialisation, and it often focused on the newly idealised childhood as the privileged lost time. Indeed, the time for which the nostalgic longs is typically childhood.

In keeping with this longstanding association, several critics have understood *The Virgin Suicides* as a classic example of nostalgia for lost childhood, with a particular focus on the innocence that childhood entails. Heusser reads *The Virgin Suicides* as 'an extended lament of [...] the loss of innocence' – the narrator's lose their youth and 'childlike unconcern'.[66] Similarly, Hoskin sees the central thematic concern of *The Virgin Suicides* as 'the destruction of childhood'.[67] However, this assumption that the nostalgia of the men in *The Virgin Suicides* is for their childhood is lazy: the men do not look back to their childhoods in general, but rather to that very particular period on the cusp of childhood and adulthood – adolescence. It is not incidental that the five sisters are aged thirteen, fourteen, fifteen, sixteen and seventeen: they span almost exactly the teenage years. Moreover, in its first few minutes the film establishes the boys as teenagers (and the accordant period of life that is being remembered as that of adolescence) when the narrator says 'even then, *as teenagers*, we tried to put the pieces together'.[68] Interestingly, this is one of very few lines of narration that is not taken directly from the novel; Coppola clearly felt a need to spell out the focus on the teenage. What, then, is the particular quality of this teenage nostalgia that the narrators of the story feel and that the audience of *The Virgin Suicides* taps into? What longings and hopes are stored therein, and how do they differ from those embodied in childhood nostalgia? Since American nostalgia films are so often focused on the teen years (*American Graffiti* and *Dazed and Confused*, exemplars of the two waves of nostalgia films, come to mind, not to mention *The Wonder Years, The Last Picture Show, Grease, My Girl, Dirty Dancing, Wet Hot American Summer* and *Stand by Me*), an exploration of teenage nostalgia in *The Virgin Suicides* might also shed some light on the nature of contemporary nostalgia culture more broadly.

One of the things that is striking in *The Virgin Suicides* is the stark contrast between the perspective of the teenagers and the adults in the film. Frequently, the adults in the film are presented as superficial in their attempted explanations of the behaviour of the teenage girls. When Cecilia, the youngest sister, first attempts suicide unsuccessfully, the narrator tells us that 'everyone had an opinion as to why', just before the camera cuts to two local women gossiping about it:

WOMAN 1: Of course the parents are to blame!
WOMAN 2: I heard they found a razor in the toilet ...
WOMAN 1: [...] That girl didn't want to die. She just wanted out of that house.
WOMAN 2: She wanted out of that decorating scheme.

The film is particularly mocking of those adults whose professions or positions lend an air of legitimacy to their attempted explanations of the girls' behaviour. The psychiatrist, Dr Horniker, played to comic effect by Danny DeVito, presents Cecilia with a Rorschach test, an immediately evocative sign of pseudoscience, given that, as a 2001 article in the *New York Times* points out, 'Almost since its creation, the inkblot test has been controversial, with early critics calling it "cultish" and later ones deeming it "scientifically useless".'[69] Cecilia smirks at him throughout the test, with a look of someone humouring an idiot. A few minutes later, Dr Horniker pronounces his verdict (wrong, it turns out) to her parents: 'I don't think Cecilia truly meant to end her life. Her act was a cry for help.'[70] He goes on to suggest that Cecilia's parents allow her to socialise more. The laughable immediacy with which Dr Horniker thinks he has understood, and diagnosed, the problem is reminiscent of Marcuse's despairing comment:

> The psychiatrist takes care of the Don Juans, Romeos, Hamlets, Fausts, as he takes care of Oedipus – he cures them. [...] the soul contains few secrets and longings which cannot be sensibly discussed, analyzed and polled.[71]

Another important figure is Ms Perl, a pushy journalist who is more interested in having a story to tell than in any effort to meet the truth on its own complicated ground. After Ms Perl conducts the briefest of 'interviews' with two of the Lisbon sisters on their porch before their mother intervenes, the film cuts to a scene in which Ms Perl appears on television, offering her summation of Cecilia's death in news bites:

> The suicide of an East Side teenager has increased the awareness of a national crisis. [...] Psychologists agree that adolescence today is much more fraught by pressures and complexities than in years past. More and more doctors say that this frustration can lead to acts of violence.[72]

The television news programme then cuts to interviews with other teenagers suffering from depression. The specificity and enormity of Cecilia's suffering is immediately slotted into place and becomes a mere data point in a web of causal explanation: 'psychologists agree', 'doctors say'.

In clear contrast with the knowing but vapid pronouncements of these various adult figures, the boys – both at the time of the narrative and in subsequent years, as they remember – are fully aware of how little they understand. Although the narrator says that 'for a while we tried to accept the general explanations, which qualified the Lisbon girls' pain as merely historic, springing from the same source as other teenage suicides, every death part of a trend', they are clearly resistant to such reductive reasoning.[73] As the narrator says near the end of the film, 'In the end, we had pieces of the puzzle, but no matter how we put them together, gaps remained, oddly shaped emptinesses mapped by what surrounded them, like countries we couldn't name.'[74] A telling comment in the novel about Ms Perl further articulates this distinction between the adult 'knowing' and the teenage desire to know: 'she cobbled together reminiscences into an airtight conclusion, far less truthful than our own, which is full of holes'.[75]

One is reminded of the adults in the film of Benjamin's *Erlebnis*, 'the mask of the adult [...] expressionless, impenetrable and ever the same the adult has already experienced everything'.[76] But in the teenage boys' unending desire to understand the lives and deaths of these mysterious sisters – riddled as their understanding is with 'gaps', 'emptinesses', 'holes' – one is reminded of Leonard Cohen's song line quoted earlier, 'There's a crack in everything/ that's how the light gets in', and of Bloch's emphasis on the 'hole in the whole', as Thompson puts it.[77] The resistance of the teenage boys to the 'official', heartless and 'airtight' so-called explanations of the girls' magical lives and mysterious deaths is a resistance to instrumental reason, to the foundational problem that Bloch describes as follows:

> It is not only easier for this age to believe in the visible than the invisible, amazingly, but even within the visible what is discrete, subdivided, seems even more real than the whole. This trend [...] leads to the total dismantling of anything original, that recognises only the mundane, the calculable, and even only the simplest impulses, contents as variable, and thus acknowledges these alone as real.[78]

Given that the film goes to some lengths to draw attention to this crucial difference between the adults and the teenagers, the notion that the film should be read as a lesson against the perils of getting stuck in the past (or as a lesson on the merits of 'growing up' in any conventional way) is uncompelling. Most adults, as the voiceover narrator reminds the audience, willingly forget, easily move on:

> We began the impossible process of trying to forget them. Our parents seemed better able to do this. Returning to their tennis foursomes and cocktail cruises. As though they'd seen this all before.[79]

Just as Benjamin says in his essay 'Experience' that 'the adult has already experienced everything', so the adults in the film have 'seen this all before'.[80] At first glance, the film might seem to suggest an undialectical opposition between teenage aliveness and imagination (*Erfahrung*) and the reductive logic that dominates adult life (*Erlebnis*): in other words, it might appear to mourn that which is necessarily lost as we grow up. However, that would be to overlook the position of the narrator(s), which is that of a voice that refuses to forget. *This* is the perspective that the viewer is invited to share.

The narrator(s) are contrasted with most of the adults presented in the film in that they have not forgotten, and cannot forget, the question of the girls' lives and deaths that consumed them as teenagers. Indeed, the fact that they are still consumed by this question is the most defining characteristic of the otherwise amorphous narrator(s). The majority of adults in the film quickly explain away the deaths of these girls as a mere statistic and return to their tennis 'foursomes', but for the teenage boys the unanswerable question of the girls' lives and deaths is, at heart, the question of life and death itself. One is reminded of Bloch's powerful claim that 'man, if he finds himself just once on the stages of this [...] ascending pyramid of ideas, *forgets the question originally motivated by self-amazement*, indeed unthinkingly allows the Encylopaedia [...] to present itself as the day of absolute maturity'.[81] If most of the adults in the film represent exactly such an encyclopedic approach to knowledge ('psychologists agree', 'doctors say'), the teenage boys, by contrast, are more in touch with the original, inconstructable 'question' of which Bloch repeatedly reminds us.

Rather than seeing the narrator(s) as in perpetual mourning and as being simply unable to move on, their perspective should be understood as a refusal to accept the trite explanations given by the world for what is fundamentally inexplicable – a refusal increasingly out of place as they have grown older and are increasingly subject to pressure to conform to the limiting version of 'adulthood' that Benjamin bemoaned in his essay 'Experience' and that the film bemoans in its representations of many of the adult characters. It is a refusal which is therefore all the more courageous.

The Virgin Suicides does not share the perspective of the 'adults' it portrays: it mocks their perspective. It draws attention to the impoverished experience of many adults precisely in order to highlight that an alternative mode of experience is possible: you remember, it seems to say to the audience, you knew it once too.

In Chapter 3 I put forward the argument that nostalgia does not always cling to a lost past, but sometimes re-engages ghosts which need redeeming, the betrayed promises of yesterday that still have relevance today. If the narrator(s) in the film are nostalgic, it is not fundamentally for the Lisbon girls as they were, but for themselves as they were. In not giving up on the girls, what they are really not giving up on is themselves. The ghost which needs redeeming, that which the men hold on to by only a bare thread in their dream-intoxicated night, that which shines only in the 'gaps' and 'holes' that have not been plugged by reductive reason, is the possibility of a richer, more whole (and hole-ier) experience: *Erfahrung*. The narrator(s) do not long for some idyllic childhood innocence, as some critics have suggested; they long for the ability they themselves once had to be awed, to wonder, to be unknowing: to meet existence without the protective, habituated, gaze of *Erlebnis*.

The teenage ideal evoked in *The Virgin Suicides* symbolises and signifies *Erfahrung*, not as something that is possible only in that particular moment of life, but as a kind of gestalt. One is reminded of Bloch's view of adolescence as the time when a foundational wonder is most acute. In a reflection in *Traces*, and again later in his Tuebingen lectures, Bloch mentions a story, *Pan*, by Norwegian novelist Knut Hamsun, in which a man meets a young adolescent girl who is suddenly filled with wonder at a simple blade of grass, and the mere fact that it is raining. Although Bloch suggests that there is a kind of amazement in childhood 'which coincides with the consciousness of being oneself – an "I"', he argues that it is only really in adolescence that 'the first true manifestation of authentic wonder is to be found in the contemplation of the ordinary itself'.[82] It is in adolescence that 'one questions more precisely, seemingly, and notices'.[83] But, although Bloch is devastated that 'almost no one kept up his questioning past the first answer' (a view that the portrayal of typical adulthood in *The Virgin Suicides* also entails) he urges that the only true knowledge is that which remains connected to that adolescent sense of wonder:

> It also ties philosophy again and again to youth, makes metaphysics at every point impatient again, conscientious – the wisdom of age in the early, unerring freshness of adolescent, primordial wonder.[84]

This, then, is the shimmer of utopian possibility that lingers in the nostalgic nightdreams of the narrator(s), demanding attention (when the adult men lie in bed 'asking God to save us', they are asking to be saved from a life experienced only as *Erlebnis*, a life that has forgotten the 'freshness of adolescent, primordial wonder'). And this is the possibility that the film, in drawing the viewer into the memories of the narrator(s) while highlighting the comparative superficiality and hypocrisy of most 'adults', wishes to remind viewers of: *Erfahrung* is the possibility it seeks to re-awaken.

Rebellion, negation and the teenage

In addition to the betrayed promise of *Erfahrung*, there is another, related, aspect of the 'teenage' experience that *The Virgin Suicides* and other films of its moment are nostalgic for: the possibility of rebellion.

Mayshark makes a distinction between *American Graffiti* (1973) and *Dazed and Confused* (1993) which is reflective of a wider distinction between the first wave of 1970s nostalgia films that looked back at the 1950s, and the second wave of 1990s nostalgia films that looked back at the 1970s. Mayshark points out that, unlike its predecessor *American Graffiti*, *Dazed and Confused* is not nostalgic for a more innocent time, but for something else:

> Post-Vietnam, post-Watts, post-Woodstock and Altamount, post-Watergate, there is no innocence left for the kids in Linklater's small-town Texas high school to lose. [...] If *American Graffiti* is nostalgic for an adolescent Eden, *Dazed and Confused* looks back longingly at that first burst of liberty out of the garden.[85]

In *American Graffiti* the teenagers are all basically happy in high school, but in *Dazed and Confused*, as one review points out, 'all the kids [...] want nothing but to be out of high school'.[86] And yet, ironically, 'all the people watching *Dazed and Confused* want nothing but to be back in high school'.[87] Similarly, in *The Virgin Suicides*, the girls are literally locked up in their house for half of the film, not to mention the broader entrapment in a bland suburbia which all the film's teenagers suffer. The teen characters in *The Virgin Suicides* yearn for an elsewhere; the girls pore over travel magazines describing faraway exotic locations and the boys buy these same magazines, following them on their imaginary journeys. Yet viewers of *The Virgin Suicides* long to be teenagers again.

The suggestion that the 1950s teenagers depicted in 1970s nostalgia films were 'basically happy in high school' is not entirely accurate. Rebellion is an important part of the representation of the Teenage even in this first wave of nostalgia films, as is clear in the famous transformation of goodie two-shoes Sandy into her leather-clad alter-ego in *Grease*. Indeed, more broadly, film critic A. O. Scott points out that the teenage rebel has long been the hero in American fiction and culture (he mentions Huckleberry Finn and Holden Caulfield, as well as the 'pouting screen rebels played by James Dean and Marlon Brando').[88]

However, it is true that in 1970s nostalgia films such as *American Graffiti* and *Grease* there is still a quality of happy innocence associated with the remembered moment (the 1950s) which is far less prominent in 1990s nostalgia films such as *Dazed and Confused* and *The Virgin Suicides*, which emphasise instead the resistance of their teen characters to their oppressive environments (their longing to get away). The nostalgia that these films thus present to their audience is less a nostalgia for lost innocence than a nostalgia for a moment of possible rebellion. An *Esquire* magazine article about *Dazed and Confused* suggests, in comic but poignant terms, that this kind of desire for rebellion may have been particularly acute in the 1990s:

> The central character Randall 'Pink' Floyd [...] doesn't want to rebel, but he doesn't want to concede to authority either. That's such a nineties predicament. I remember it: were you going to sell out? Were you going to join the man? Everybody did sell out, of course, but everybody also wanted to preserve the aura of outrage, the feeling of alternatives. It is the single most annoying feature of Generation X – tenured professors who believe they are countercultural radicals, lawyers who think they're rebels because they know about hip-hop, accountants who 'keep it real.' [...] That mixture of faux anti-authoritarianism and intensely self-obsessed nostalgia makes *Dazed and Confused* the definitive film of Generation X.[89]

Rather than this just being a characteristic quirk of Generation X, I want to suggest that the heightened desire for rebellion embodied in nostalgia films of the 1990s and since, in fact reflects a deeper struggle with questions of authority and rebellion in the post Cold War landscape. In particular, it is reflective of the increasing difficulty of any kind of negation in contemporary society (see Interlude 2). In basic terms, with the advancement of a fluid information society, and in the absence of the Cold War Manichean moral structure, it has become harder to identify the 'baddies' against whom one might rebel. If during an earlier period of capitalism the agents of production were more visible (the owners of capital standing overhead the factory floor, to over-simplify the picture), under late capitalism, those same agents are everywhere and nowhere, or so it seems. Within American society, Marcuse already in the late 1950s had criticised the increasing assimilation of oppositional culture into mainstream culture; and by the time of young Tavi the lack of counter-culture among her peers has seemingly solidified into an irrefutable reality: 'there used to be hippies and beatniks, now there's nothing!' Tavi laments.[90] Is it any wonder, in this context, that a cultural nostalgia for the Teenage – and the possibility of rebellion, negation or opposition that the teenage ideal signifies – should heighten?

Scott suggests that although the teenage figure has long been prominent in American fictions, the rebellion of these antiheroes of the past 'still accepted the fact of adulthood as its premise'.[91] He goes on to discuss the increasing prominence of young adult fiction (a third of readers of young adult fiction today are adults), the prevalence of comic man-boy eternal adolescents in films like *The Hangover* (2009) and *The 40-Year-Old Virgin* (2005), and the death of Don Draper in *Mad Men* (the last American adult?), and concludes by claiming that there is a crisis of adulthood in American culture. As he points out:

> In my main line of work as a film critic, I have watched over the past 15 years as the studios committed their vast financial and imaginative resources to

> the cultivation of franchises [...] that advance an essentially juvenile vision of the world. Comic-book movies, family-friendly animated adventures, tales of adolescent adventures and comedies of arrested development do not only make up the commercial center of 21st-century Hollywood. They are its artistic heart. [...] What all of these shows grasp at, in one way or another, is that nobody knows how to be a grown-up anymore. Adulthood as we have known it has become conceptually untenable.[92]

The nostalgia for adolescence typified in films like *The Virgin Suicides* and *Dazed and Confused* could be perceived as the other side of this crisis of adulthood. After all, adolescence requires adulthood to make any sense of itself. Marcuse's moment of liberation, of negation, requires a primordial father against whom the brother clan rebels. The man-children of films like *The Hangover* seem, in their pathetic 'arrested development', to be a visual embodiment of the tyranny of tolerance that Jameson describes and of the fact that today's consumer is 'denied that path towards genuine psychic individuality once offered him by the revolt against the father'.[93] If all oppositional culture is incorporated into the mainstream, then a so-called crisis of adulthood is perhaps equally a crisis of the Teenage. The man-children of *The Hangover* can, on the surface at least, do whatever they like, for as long as they like, and their resultant impotence is evident. The teenagers in *The Virgin Suicides*, by contrast, are kept under lock and key, but develop their own erotic and desirous language, their own realm, decidedly distinct from the impoverished adult world of cocktail parties and tennis foursomes that surrounds them; it is a realm in which they have a relative freedom – of thought if not action.

If nostalgia for childhood arose in resistance to the speed of industrialisation and contained within it a forgotten utopian promise, the promise of nature and innocence, the teenage nostalgia prevalent in American nostalgia films from the 1990s onwards can be seen to express a muffled desire for the possibility of rebellion and negation, the actual levers of which seem in reality hard to locate given the diffuseness of power in late capitalism and the accordant sense that 'there is no alternative'.

This should not be understood as a confirmation of hopelessness; the nostalgia does not express a desire for that which is by definition past, and therefore necessarily no longer possible. As explored in depth in Chapter 3, nostalgia is often a desire to re-engage the *still not*, the *Not Yet* fulfilled, but not quite forgotten, potentials of yesterday. In this case, the desire for rebellion that nostalgia films such as *The Virgin Suicides* and *Dazed and Confused* exhibit could be seen as the shimmer, or spectre, of a forgotten but not yet dead possibility: the possibility of negation. As the analysis in Chapter 3 highlighted, the memory of negation, of the moment of possible liberation by the brother clan, is an important component of any utopianism: it reminds us that things need not be as they are. The strong desire for the possibility of negation that these films reflect and feed into is an indication that there is dissatisfaction with the incorporation of all 'alternative' cultures in late capitalism. The very desire for negation indicates a degree of resistance to the current situation, even if it is a muffled resistance, one that cannot yet name itself and does not yet know what its real avenues are.

In summary, then, in the proliferation of teenage nostalgia in films like *The Virgin Suicides* lies a latent spark of resistance to a society in which *Erfahrung* has given way to *Erlebnis* and the possibility of negation (conceptual or actual alternatives to 'what is') is frustrated. These are the shimmering utopian promises that continue to insist, haunting the contemporary imaginary.

Notes

1 Jameson, '*The Shining*', p. 116.
2 Sprengler, p. 3.
3 Jameson, 'Postmodernism' (1984), p. 67.
4 Gleiberman, 'Why Nostalgia Movies Leave us Dazed (but Not Confused)'.
5 Sprengler, p. 45.
6 Jesse Fox Mayshark, *Post-pop Cinema: The Search for Meaning in New American Film* (Westport, CT: Praeger, 2007), p. 22.
7 Potter, 'Mad Men and the end of Delicate, but Potent Nostalgia'.
8 Sprengler, p. 1.
9 The focus in this chapter is on *The Virgin Suicides* film, considered alongside other nostalgia films. However, many passages of narration and dialogue are taken directly from the book, and Coppola's loyalty to Eugenides' novel, particularly to the tone of his work, has been noted by several critics. For this reason, I will also look to the novel where it helps to shed light on the intention or spirit of the film.
10 Peter Guttridge, 'Book Review: Dying to Escape Big Bad Mum: *The Virgin Suicides* – Jeffrey Eugenides', *Independent*, 18 June 1993, www.independent.co.uk/arts-entertainment/books/book-review-dying-to-escape-big-bad-mum-the-virgin-suicides-jeffrey-eugenides-bloomsbury-1599-pounds-1492519.html [accessed 5 May 2015].
11 Dines, p. 63.
12 Ed Gonzales, '*The Virgin Suicides*', *Slant Magazine*, 2 May 2001, www.slantmagazine.com/film/review/the-virgin-suicides [accessed 6 May 2015].
13 Kharissa Forte, 'Cult Classic Review: *Virgin Suicides* Presents an Eerie Nostalgia', *University of Missouri, Kansas City, University News*, 24 September 2014, http://info.umkc.edu/unews/cult-classic-review-virgin-suicides-presents-an-eerie-nostalgia/ [accessed 13 May 2015].
14 Keith Phipps and Tasha Robinson, 'Forum: *The Virgin Suicides*', *The Dissolve*, 25 June 2015, https://thedissolve.com/features/movie-of-the-week/1080-forum-the-virgin-suicides/ [accessed 25 June 2015].
15 Martin Heusser, 'Et in Arcadia Ego: The Pastoral Aesthetics of Suburbia in Jeffrey Eugenides' *The Virgin Suicides*', *Swiss Papers in English Language and Literature*, 20 (2007), pp. 175–189 (p. 175).
16 Nick Jolly, '*The Virgin Suicides* and the Transcendental Quietude of Young Love', *The Essential*, http://theessential.com.au/features/nostalgia-files/the-virgin-suicides-and-the-transcendental-quietude-of-young-love [accessed 12 February 2016].

17 Peter Cox, 'Why I Love … #56: *The Virgin Suicides*', *The Lost Highway Hotel*, 11 October 2013, https://thelosthighwayhotel.com/2013/10/11/why-i-love-56-the-virgin-suicides/ [accessed 6 May 2015].
18 Gevinson, 'Moving Images: Shows and Films Imbued with Yearning'.
19 TIME Staff, 'The 25 Most Influential Teens of 2014', *Time Magazine*, 13 October 2014, http://time.com/3486048/most-influential-teens-2014/ [accessed 7 May 2015].
20 Tavi Gevinson, 'Girls with Power and Mystique: An Interview with Sofia Coppola', www.rookiemag.com/2013/06/sofia-coppola-interview/ [accessed 8 May 2015].
21 Sprengler, p. 83.
22 Jameson, *Postmodernism* (1992), p. 18.
23 Jameson, '*The Shining*', p. 114.
24 Mark Fisher, 'Corridors in which Time is Abolished: Notes on Last Year in Marienbad', *K-Punk*, 4 October 2005, http://k-punk.abstractdynamics.org/archives/2005_10.html [accessed 7 May 2015].
25 Jameson, '*The Shining*', p. 116.
26 Jameson, '*The Shining*', p. 116.
27 Jameson, '*The Shining*', p. 116.
28 Beuka, p. 4.
29 Beuka, p. 10.
30 Coon, p. 68.
31 Phipps and Robinson, 'Forum: *The Virgin Suicides*'.
32 Bree Hoskin, 'Playground Love: Landscape and Longing in Sofia Coppola's *The Virgin Suicides*', *Literature/Film Quarterly*, 35.3 (2007), 214–221 (p. 215).
33 Hoskin, p. 216.
34 Gonzales, '*The Virgin Suicides*'.
35 Melissa Hunter, '*The Virgin Suicides*', *The Soul of the Plot*, 12 May 2015, https://thesouloftheplot.wordpress.com/2015/05/12/the-virgin-suicides/ [accessed 13 May 2015].
36 Peter Preston, 'Death and the Maidens', *Guardian*, 21 May 2000, www.theguardian.com/film/2000/may/21/peterpreston [accessed 8 May 2015].
37 R. Colin Tait, 'That 70s Sequence: Remembering the Bad Old Days in *Summer of Sam*', *Cinephile*, 5.2 (2009) http://cinephile.ca/archives/volume-5-no-2-the-scene/that-70s-sequence-remembering-the-bad-old-days-in-summer-of-sam/ [accessed 2 March 2016].
38 Tait, 'That 70s Sequence'.
39 Tait, 'That 70s Sequence'.
40 Tait, 'That 70s Sequence'.
41 Tait, 'That 70s Sequence'.
42 Mayshark, pp. 170–171.
43 Tait, 'That 70s Sequence'.
44 Mayshark, pp. 1–2.
45 Claire Marie Healey, 'Deconstructing the Fashion of *The Virgin Suicides*', *Dazed*, 2015, www.dazeddigital.com/fashion/article/24730/1/deconstructing-the-fashion-of-the-virgin-suicides [accessed 20 August 2016].
46 Both are easily accessible online but it was not possible to get permission to reprint the Marc Jacobs *Daisy* campaign here.
47 Jameson, '*The Shining*', p. 117.
48 Jameson, '*The Shining*', p. 123.
49 Jameson, '*The Shining*', p. 123.

50 Fredric Jameson and Anders Stefanson, 'Regarding Postmodernism – A Conversation with Fredric Jameson', *Social Text*, 21 (1989), 3–30 (p. 11).
51 Coon, pp. 50–56.
52 Coon, p. 64.
53 Dika, p. 89.
54 Dika, p. 94.
55 Dika, p. 94.
56 Matt Singer, 'Keynote: *American Graffiti* and George Lucas' Nostalgia for Nostalgia', *The Dissolve*, 18 March 2014, https://thedissolve.com/features/movie-of-the-week/468-keynote-american-graffiti-and-george-lucas-nostalg/ [accessed 9 May 2015].
57 Gleiberman, 'Why Nostalgia Movies Leave us Dazed (but Not Confused)'.
58 Dines, p. 963.
59 Debra Shostak, '"A Story We Could Live With": Narrative Voice, the Reader, and Jeffrey Eugenides' *The Virgin Suicides*', *Modern Fiction Studies*, 55.4 (2009), 808–832 (p. 813).
60 *The Virgin Suicides*, dir. by Sofia Coppola (Paramount Classics, 1999).
61 *The Virgin Suicides*, dir. by Sofia Coppola.
62 Jeffrey Eugenides, *The Virgin Suicides* (London: Fourth Estate, 2013), p. 142.
63 Coon, p. 64.
64 Žižek, Preface, *The Privatization of Hope*, p. xix.
65 Jenß, p. 113.
66 Heusser, p. 178.
67 Hoskin, p. 214.
68 *The Virgin Suicides*, dir. by Sofia Coppola (my italics).
69 Erica Goode, 'What's In an Inkblot? Some Say, Not Much', *New York Times*, 20 February 2001, http://mobile.nytimes.com/2001/02/20/science/what-s-in-an-inkblot-some-say-not-much.html?_r=0 [accessed 3 July 2016].
70 *The Virgin Suicides*, dir. by Sofia Coppola.
71 Marcuse, *One-Dimensional Man*, p. 74.
72 *The Virgin Suicides*, dir. by Sofia Coppola.
73 Eugenides, *The Virgin Suicides*, pp. 232–233.
74 *The Virgin Suicides*, dir. by Sofia Coppola.
75 Eugenides, *The Virgin Suicides*, p. 217.
76 Benjamin, 'Experience', *Selected Writings*, p. 3.
77 Thompson, 'The Frankfurt School, Part 2: Negative Dialectics'.
78 Bloch, *The Spirit of Utopia*, p. 166.
79 *The Virgin Suicides*, dir. by Sofia Coppola.
80 Benjamin, 'Experience', *Selected Writings*, p. 3.
81 Bloch, *The Spirit of Utopia*, p. 197 (my italics).
82 Bloch, *A Philosophy of the Future*, p. 5.
83 Ernst Bloch, *Traces*, trans. by Anthony A. Nassar (Stanford: Stanford University Press, 2006), p. 170.
84 Bloch, *Traces*, p. 170.
85 Mayshark. p. 23.
86 Marche, '*Dazed and Confused* was the Definitive Movie About the '90s, Not the '70s'.
87 Marche, '*Dazed and Confused* was the Definitive Movie About the '90s, Not the '70s'.
88 A. O. Scott, 'The Death of Adulthood in American Culture', *New York Times*, 11 September 2014, www.nytimes.com/2014/09/14/magazine/the-death-of-adulthood-in-american-culture.html [accessed 5 May 2015].

89 Marche, '*Dazed and Confused* was the Definitive Movie About the '90s, Not the '70s'.
90 Dan Casey, 'Rookie of the Year, Again: An Interview with Tavi Gevinson', *Nerdist*, 5 December 2013, http://nerdist.com/rookie-of-the-year-again-an-interview-with-tavi-gevinson/ [accessed 6 May 2015].
91 Scott, 'The Death of Adulthood in American Culture'.
92 Scott, 'The Death of Adulthood in American Culture'.
93 Jameson, *Marxism and Form*, p. 109.

Interlude 3

Marshalling the past: utopia versus once upon a time

Throughout his life and work, Benjamin was highly critical of the 'worship of progress', often articulating a 'deep distrust of the ideology of progress'.[1] He saw the 'smug complacent affirmation of progress' as 'the quintessential modern myth' that needed exploding.[2] This criticism is particularly strong in Benjamin's 'Theses on the Philosophy of History', in which he argues that:

> The concept of historical progress of mankind cannot be sundered from the concept of its progression through a homogenous, empty time.[3]

His criticism of the narrative of progress is that it tells a story – much like many children's books – which assumes an already decided arc. The reader of children's fairy-tales knows that the 'once upon a time' will end in 'happily ever after', just as the complacent teller of the progress-myth 'knows' that everything can be made sense of within the framework of inevitable progress. This is a charge that Benjamin levelled not only at the predominant liberal version of progress, but equally at any Marxism that assumes that history will necessarily progress dialectically, as though the question of how man responds to history were not urgent. He saw the role of the true historian – of the historical materialist, that is – to challenge such smug complacency:

> The historical materialist leaves it to others to be drained by the whore called 'Once upon a time' in historicism's bordello. He remains in control of his powers, man enough to blast open the continuum of history.[4]

Bloch, due to his emphasis on the coming utopian 'Heimat' and his descriptions of the tendency-latencies emergent in all things, is often mistaken as a pedlar of such a progress story. David Kaufman, for example, in 'Thanks for the Memory: Bloch, Benjamin and the Philosophy of History', contrasts Benjamin's vigorous critique of progress with what he sees in Bloch as an 'apparent faith to an outmoded philosophy of history'.[5] However, for Bloch, these tendency-latencies are never guaranteed, but only ever possibilities, which rely on exhaustive effort and which require human agents to rise to the challenge, to respond to the 'inconstructable question' that existence asks of us, and bring forth into reality what is otherwise only potential.

This is clearest in an often overlooked, but hugely powerful, section of *The Spirit of Utopia*, called 'Kant and Hegel, or, Inwardness Overtaking the Encyclopedia', in which Bloch defends Kant against Hegel in one important respect. In this section, Bloch criticises Hegel's certainty that the movement of history is ultimately toward Absolute Spirit, and argues that there is too much reassurance in the Hegelian 'panlogical pathos of perfection', too much certainty that 'the secretary of the world-spirit triumphs in an unsubjective, panlogical objectivity'.[6] He insists that this certainty devalues the urgent moral fight of the experiencing subject, the weight of responsibility that falls on humans to live into and create their own futures. What Bloch values most highly in Kant is the inexhaustibility, urgency, humanity and moral responsibility that the Categorical Imperative poses:

> In Kant, in other words, philosophy was a solitary light meant to burn up the night of this world. In Hegel philosophy becomes a headmaster, or indiscriminate lawyer for the Being that hired him, and the night of the world retreats into the merely ignorant subject. Here spreads the beautiful warmth of the classroom, so that everything painful, unendurable and unjust about life, the constant necessity of its refutation, the self-immolation of nature and the entire Herculean passion of the idea, can be developed into something safe, always occurring, never occurring, whose proper exposition is either just written on the blackboard, or else, in accordance with the eternally resolved, eternally completed logological silence of actual reality, is a mere ceremony.[7]

In the urgency that Bloch accords to the ethical call issued to us by tomorrow, he values not only Kant but also an existentialist understanding of freedom, stating for example that 'it is far preferable, as Kierkegaard says, to be angry with Jesus, yet constantly in relationship to him, than to be a spectator who has understood Him'.[8]

Far from being a proponent of a grand teleology (an accusation sometimes made against him), these passages show that Bloch, like Benjamin, is highly critical of any narrative of pre-ordained progress. For both Bloch and Benjamin, part of the value of bringing the past into rupturing dialogue with the present and future, has to do with challenging a notion of linear progress. Kaufman says of Benjamin that:

> By making the present the custodian of the past, Benjamin asserts that history will not take care of itself. Oppression will not cancel itself out as if the world were studiously dialectical.[9]

In a similar vein, Bloch's conception of the dialectic is explicitly contrasted with a too-predictable dialectic, one which is perceived as always necessarily pushing forward at the front of history. For Bloch, it is not only contemporaneous contradictions (of class interests, for example) that push history forward; non-contemporaneous contradictions can also result in the emergence of something new. He asks that the dialectic be understood as a 'multi-layered revolutionary dialectic; for obviously the entirety of earlier development is not yet "resolved" in capitalism and its dialectic'.[10] This idea of a 'multi-temporal and multi-spatial dialectic' is an important aspect of Bloch's thought.[11] Kellner explains how Bloch's view of the unfolding or materialisation of history is far from linear, and how Bloch's conception of the latent utopian promise within non-contemporaneous material makes sense within his conception of history:

> Bloch urges us to grasp the three dimensions of human temporality: he offers us a dialectical analysis of the *past* which illuminates the *present* and can direct us to a better *future*. The past – what has been – contains both the sufferings, tragedies and failures of humanity – what to avoid and what to redeem – and its unrealized hopes and potentials – which could have been and yet can be. For Bloch, history is a repository of possibilities that are living options for future action; therefore what could have been can still be. [...]
>
> This three-dimensional temporality must be grasped and activated by an anticipatory consciousness that at once perceives the unrealized emancipatory potential in the past, the latencies and tendencies of the present, and the realizable hopes of the future.[12]

Thus we arrive at a final aspect of the utopian, or rather of the past-utopian. Just as 'still not' fulfilled promises can re-awaken contemporary (wo)man, making her aware of what is *still* missing, so these lingering ghosts can also jolt her out of the complacent and dangerous assumption that the future will automatically unfold in a certain way, reminding her of the urgent fight that is still necessary for the ethical subject. Utopia is not, for Bloch, something that mankind necessarily inches towards: it is only ever a possibility on the horizon that demands our urgent engagement. As such, the unfinished utopian promises of yesterday can function to highlight that need, and the risks involved in complacency.

Notes

1 Kaufman, p. 45, p. 43.
2 Gilloch, p. 129.
3 Benjamin, Walter, 'Theses on the Philosophy of History', in *Illuminations: Essays and Reflections*, trans. by Harry Zohn (New York: Schocken Books, 1969), pp. 253–265 (p. 261).
4 Benjamin, 'Theses on the Philosophy of History', *Illuminations*, p. 262.
5 Kaufman, p. 43.
6 Bloch, *The Spirit of Utopia*, p. 181.
7 Bloch, *The Spirit of Utopia*, p. 185.
8 Bloch, *The Spirit of Utopia*, p. 180.
9 Kaufman, p. 46.
10 Bloch, *Heritage of Our Times*, p. 114.
11 Bloch, *Heritage of Our Times*, p. 115.
12 Kellner, p. 81.

6 Nostalgia in photographs of Detroit's 'abandoned spaces'

In his *Paris Review* interview, Jeffrey Eugenides speaks about the significance of his childhood home in the Detroit suburbs:

> My entire childhood coincided with the demise of Detroit. I grew up watching houses and buildings fall apart and then disappear. It imbued my sense of the world with a strong elegiac quality – a direct experience of the fragility and evanescence of the material world.[1]

That elegiac quality permeates both his novel and Coppola's adaptation. Although the decline of the motor industry is only once referenced in *The Virgin Suicides*, the film is throughout imbued with a sense of impermanence, of something slipping away (whether the teenage years or a time of American self-assurance and prosperity), a feeling that is enhanced by the film's unremarked backdrop. This implicit background of *The Virgin Suicides* – the decline of Detroit – is the focus of the work considered in this chapter: photography of Detroit's urban ruin.

Contemporary fascination with 'abandoned spaces', sometimes also referred to by commentators as 'abandoned places' or 'industrial ruin', is a phenomenon that has been widely observed. DeSilvey and Edensor discuss an 'extraordinary intensification of academic and popular interest in the ruins of the recent past', arguing that 'we seem to be in the midst of a contemporary *Ruinenlust*'.[2] Huyssen analyses the 'contemporary obsession with ruins', which he calls 'a contemporary nostalgia [...] for the decaying residues of the industrial age and its shrinking cities'.[3] Strangleman takes as his starting point a widespread 'smokestack nostalgia'.[4] And Hell and Schonle's book of collected essays titled *Ruins of Modernity* begins with the observation that although America used to show little interest in its own ruins, 'in a post-Fordist, post 9/11 reality, the imaginary of imperial ruin and ruination has become pervasive'.[5]

The increasing interest in 'abandoned spaces' is particularly evident in the proliferation, in the last decade, of coffee table photography books consisting of images of urban and industrial ruin, and online spaces devoted to the same photographic subject. The trend begins most obviously with Camilo Vergara's *American Ruins* (1999), but subsequent well-known publications include (but are not limited to) Sylvain Margaine's *Forbidden Places: Exploring our Abandoned Heritage* (2009), Andrew Moore's *Detroit Disassembled* (2010), Yves Marchand and Romain Meffre's *The Ruins of Detroit* (2010), Dan Dubowitz's *Wastelands* (2010), Tong Lam's *Abandoned Futures: A Journey to the Posthuman World* (2013), Hank Van Rensbergen's *Abandoned Places: The Photographer's Selection* (2014), Matthew Christopher's *Abandoned America: The Age of Consequences* (2014), Seph Lawless's *Autopsy of America* (2014) and Richard Happer's *Abandoned Places: 60 Stories of Places Where Time Stopped* (2015). Nearly all of these photographers also have websites, many of which are devoted exclusively to 'abandoned spaces', such as Matthew Christopher's www.abandonedamerica.us and Hank Van Rensbergen's www.abandoned-places.com. Other photographers have become well known for their 'abandoned spaces' photography online, although they have not yet published photography books. For example, Christian Richter (www.richterchristian.com) appears in numerous popular press articles and has 14.2k followers on Instagram, and Tag Christof's Instagram profile 'americaisdead' has 36.9k followers. Even filmmakers have been entering the fray: David Lynch, for example, has produced *The Factory Photographs* (2014), a collection of previously unpublished photographs, taken over the preceding decades, of derelict factories in Berlin, Lodz, New York, New Jersey and England.

This chapter focuses on photography of Detroit's 'abandoned spaces' because Detroit is in many ways the quintessential locus for this genre of photography, as is clear from the journalistic commentary about the genre. Leary points out in his article 'Detroitism' that although photographers, ruin websites and Urban Exploration blogs chronicle ruin across North America and Europe, 'Detroit remains the Mecca of urban ruins'.[6] Other locations, particularly in ex-Soviet countries, have experienced a recent surge in ruin fascination but, as one commentator on ex-Soviet ruin photography points out, even these newly popular geographies of abandonment still exert less pull than Detroit, 'the one competitor in the 21st century ruin market that – despite their exoticism and militaristic chic – the far-flung corners of Terra Sovietica must cede to'.[7]

The central question here, mirroring those asked in previous chapters, is: Does this photographic genre peddle an aestheticised consumable nostalgia that effaces history, or does the nostalgic longing reflected in these images contain critical or utopian potential?

I will focus my analysis on four coffee table photography books, each by a different photographer. The first is *American Ruins* (1999), by Camilo José Vergara, a renowned international photographer whose images of Detroit's ruins first began to make them famous.[8] The second is *The Ruins of Detroit* (2010) by French photographers Yves Marchand and Romain Meffre, also well-known professional photographers in this domain.[9] The third is *Abandoned Places* (2007) by Henk Van Rensbergen, an airline pilot and amateur photographer with a passion for photographing abandoned spaces around the world who has published several photography books on the subject.[10] The fourth is *Abandoned Futures: A Journey to the Posthuman World* (2013) by Tong Lam, a history professor at the University of Toronto who also works as a visual artist, documenting various contemporary ruins in order to ask whether it is 'possible to rethink the relation between politics and aesthetics through infrastructures and ruins'.[11]

My method combines analysis of the photographic images and close reading of the written text in these photography books, as well as analyses of how these images are consumed by contemporary readers and of whether these patterns of consumption affect the nature of the impact of these books and images.[12] Although my focus is on Detroit, my goal is also to shed light on the broad character of this genre of 'abandoned spaces' photography.

Detroit 'abandoned spaces' photography: the current debate

The reason that Detroit has so attracted urban explorers and photographers of abandonment is obvious: as Steinmetz points out, 'Detroit is widely known as the most impoverished and abandoned major city in the United States.'[13] Having once symbolised prosperity and industry, the infamous 1967 riots exposed a political and racial crisis in the city and foreshadowed the gruelling decline which it would endure over the following decades. The decline of the motor industry in the 1970s and 1980s, combined with mass suburbanisation, left a city that was – statistically speaking – a shadow of its former glory. The population in 2010 was 713,777, a 60 per cent drop from its peak population of over 1.8 million, recorded in the 1950 census.[14] A city that had once been brightly illuminated by the lights of the flourishing car industry came to signify something very different in the American cultural imaginary. Dora Apel puts it starkly: the former Motor City has become the 'poster child of ruination in the advanced capitalist countries today'.[15] Atkinson and Clayton likewise argue that 'in Detroit, the dominant cityscape is often ordered around the disorder, danger, neglect and physical ruins of the city'.[16] The city is now home to thousands of derelict buildings, large and small, industrial and residential. Most famously, the former Michigan Central Station has stood empty since 1988, un-renovated and imposing, an image of decay that countless photographers have captured.

There is already a lively debate surrounding 'abandoned spaces' photography of Detroit. Edensor and DeSilvey describe a recurrent tension among theorists of modern ruins 'between assertions about the potential for ruins to challenge and critique normative social and material formations and counter-claims about their potential associations with regressive politics and aestheticized passivity'.[17] The critical voices are often louder. 'Ruin porn', as it is damningly dubbed, is criticised by local residents, journalists and theorists for aestheticising poverty and ignoring the very real social reality. As the title of one magazine article, a response to Marchand and Meffre's work, scathingly puts it, 'poverty is so photogenic'.[18] Residents are critical of the mainly white middle-class visitors to their abandoned hometown, and, as another article comments, often speak out 'against the artists who drop in on writing assignments and photography projects to turn the image of a city's pain into a curiosity'.[19] In addition to these out-of-towners, Steinmetz writes about the 'steady stream of suburban visitors' and the fact – pointed out with disdain – that 'bookstores in Detroit's suburbs carry shelves of paperbacks with sepia-toned covers published by Arcadia, a company that describes itself as being "pretty much all nostalgia"'.[20]

On the other side of the argument, several theorists and critics, not to mention the photographers themselves (for whom the characterisation of their varied projects as 'ruin porn' no doubt grates), have sought to defend 'abandoned spaces' photography against these accusations. Strangleman offers 'a critical reading which nonetheless attempts to find a more positive account of what these manifestations of nostalgia for an industrial past might mean'.[21] Blackmar, likewise, looks at four photographers of American ruins and argues that nostalgia is not always an ideological or uncritical sentiment, but can be used to critical effect.[22] These efforts align broadly with my own, although, as I will explain, they are differently oriented.

In the two previous chapters I have followed the same approach: a reading of the nostalgia text in question as exemplary of Jameson's 'nostalgia mode' ('the rubble') and a subsequent Bloch-inspired reading which looks for critical or utopian potential within the same text ('the gold in the rubble'). The latter readings do not invalidate the former but do suggest that they are insufficient: that utopia permeates ideology and that latent elements of critical and utopian potential are contained within what might at first glance appear to be superficial and ideologically driven texts. My Blochian readings (so termed because they make use of Bloch's broad critical approach to popular culture and his views on non-contemporaneity), also drew on Benjamin and Marcuse, theorists that I grouped together with Bloch as useful for approaching contemporary forms of nostalgia in a nuanced way.

In this chapter, however, I will pursue a slightly different approach, separating out a Benjaminian reading from a Blochian reading. This is not to undermine the contention that I have hitherto maintained, which is that both thinkers can be drawn on to resist any easy rejection of contemporary nostalgia texts and are in this regard aligned. But the aim here is to draw out differences in approach between Benjamin and Bloch that are relevant for present purposes. The contours of the *existing* debate can broadly be categorised as 'Benjaminian' versus 'Jamesonian', and I hope that my discussion of each perspective, and the readings I give in support of each perspective, will add depth to that debate. But I aim also to highlight a particular problem facing the existing, largely Benjamin-inspired defences of ruin photography, one that I believe a Bloch-inspired approach can help us to think through.

A Benjaminian reading of Camilo José Vergara's *American Ruins*

As discussed in Chapter 3, Benjamin, like Bloch, sought to redeem the ruins of the past, to liberate from within them what may yet have unfinished utopian potential. Given Benjamin's extensive work on the Paris arcades, which, though not actually abandoned, were increasingly obsolete, it is not surprising that he has often been used to theorise contemporary ruin. As Edensor and DeSilvey point out, many attempts to read contemporary ruins as possible sites of critical resistance are influenced by, or rely on, the work of Walter Benjamin.[23]

One example of this is the exploration of 'Mundane Hauntings: Commuting through the Phantasmagoric Working-Class Spaces of Manchester, England' by Edensor.[24] He describes the contemporary city as a 'palimpsest composed of different temporal elements' and argues that the 'modern imperative to swiftly bury the past produces cities that are haunted by that which has been consigned to irrelevance'.[25] He writes of his commute to work in Manchester that 'traces of the past linger in mundane spaces' and 'burst forth [...] like slips of the tongue', clearly evoking Benjamin's application of Proust's *mémoire involontaire* in the cityscape, where encountering fragments from the past can shock us out of our complicity with the present, allowing for moments of remembered *Erfahrung* as opposed to the *Erlebnis* which dominates our daily experience. Edensor, like Benjamin, sees the unfulfilled utopian potential of shadow pasts that continue to haunt the landscape. For example, in ex-council estates he describes how 'some standardized features remain in doors, windows, roofing and brick facades', which he understands as communal living's 'refusal to disappear', which 'haunts attempts to personalise property'.[26] In other words, traces of the lost post-war promises of collective living and state support can still be seen in the now abandoned or re-functioned housing projects, reminders of forgotten hopes of social justice that could still be re-awakened.

Another example of a Benjaminian reading of modern ruins is Justin Armstrong's ethnographic fieldwork in abandoned small towns in rural Saskatchewan, Canada, for which he explicitly draws on Benjamin's figure of the flâneur (or urban wanderer) to describe his experience as a rural flâneur. His initial observation is that in small towns across Saskatchewan 'there is a build-up of history written in the ruination and collapse of abandoned settlements'.[27] He distinguishes Benjamin's urban flâneur (who 'seeks refuge in a crowd') from the rural flâneur (or, more precisely, the flâneur of abandonment) who 'finds solace in the collection of memories, embedded histories and the material remnants of lives-once-lived'.[28] But in both, the unplanned nature of the wandering is crucial as it allows for random encounters both with people and the environment, and for the possibility that the modern subject might be awakened from the dream house that he inhabits (which is navigated always within the framework of a habituated and systematic consciousness), by glimpses into other no-longer or not-yet habitual possibilities. It is the unhabituated, or un-utilised, nature of abandoned spaces that, according to Armstrong, allows the flâneur to be touched or awakened by them, and by their truth.

Both Armstrong and Edensor draw on a number of Benjamin's ideas, but most central among these is the notion that the city (or town) consists of layers of history. In Convolute C of *The Arcades Project* Benjamin writes:

> Paris is a counterpart in the social order to what Vesuvius is in the geographic order: a menacing, hazardous massif, an ever-active hotbed of revolution. But just as the slopes of Vesuvius, thanks to the layers of lava that cover them, have been transformed into paradisal orchards, so the lava of revolutions provide uniquely fertile ground for the blossoming of art, festivity, fashion.[29]

The layers of volcanic lava are in this passage likened to the layers of history on top of which the phantasmagoria of the modern world have been built. But these layers, which remain embedded in the city, contain forgotten utopian, revolutionary potential. The comparison with Vesuvius suggests that the layers of history that make up a city, although normally hardened and serving as a bedrock on which the modern phantasmagoria unconsciously thrives, can still come to life (as if loosened into flowing lava again, if only momentarily) in moments or jolts of recognition. For Benjamin, each of these layers provides its own clues and dormant promises. The arcades are understood as 'galleries leading into the city's past'.[30] The metro too: 'at dusk glowing red lights point the way into the underworld of names' that 'have all thrown off the humiliating fetters of street or square, and here in the lighting-scored, whistle-resounding darkness are transformed into misshapen sewer Gods, catacomb fairies'.[31]

But if Benjamin's ideas have been borrowed to theorise the critical potential of actual abandoned spaces as challenging the hegemonic organisation or habituated experience of the cities and towns in which we live and move, can photographic imagery of Detroit's industrial ruin be understood in a similar way? Can the evocative reminders and awakenings that Edensor and Armstrong experience in these layered landscapes, in driving through Manchester or walking through small-town Canada, function similarly via photographs?

To answer these questions, I turn to the work of Camilo José Vergara. Born in 1944 in Chile, Vergara is a New York-based photographer whose photography of American ruins, slums and abandoned spaces began in the 1980s and continues to this day. He has authored eight photography books (in which images are accompanied by essayistic reflections), of which the best known and most exemplary instance of ruin photography is *American Ruins* (1999). His feeling regarding the ruination that he documents is elegiac and mired in aesthetic appreciation: 'I record urban decay with a combined sense of respect, loss, and admiration for its peculiar beauty.'[32]

Vergara sees his work first and foremost as a kind of witnessing or testimony. He repeatedly calls his photographic practice 'documentation' and is known for his diligent and committed methods. Almost every image in *American*

Ruins is accompanied by a historical explanation of the site in the image and the story of its abandonment, reflections about its particular significance and anecdotes from interviews with local residents. For example, alongside several photographs of the old Firemen's Insurance Headquarters in Newark, Vergara writes:

> I am fascinated by the entropic remains of the American city's golden age, by those 'splendid,' 'arrogant,' and 'worldly' buildings that were meant to herald 'the new, the coming thing.' In the 1920s these grand buildings captured the imaginations of Erich Mendelsohn, Fritz Lang, and so many other European visitors. They symbolized America. And such grand creations continue to radiate their tremendous energy, even after being abandoned. My mission has been to account for their recent past and to articulate their meanings. I want to pay homage to their faded dreams.[33]

Vergara believes he is documenting these 'faded dreams', the layers of bygone promise built into the material fabric of the city itself. He witnesses (and thereby refuses to forget) the remnants of past utopian energies still somehow radiating in these now crumbling structures. Lewandowsky argues that Vergara's work, like Benjamin's, performs a 'rescuing critique'.[34] Although faced with different geopolitical realities, he suggests, both Benjamin and Vergara resist the predominant narrative of 'progress', which both of them see as amounting 'to a kind of rage against the particular, the unassimilated and the contingent of social life'.[35] By re-visiting or rescuing the 'refuse of history' Vergara, like Benjamin, refuses to allow the different stories and potentials contained therein to be discarded or forgotten in the name of progress. Lewandowsky calls this an 'anamnestic obligation' to an urban past: 'What fascist and neo-liberal "progress" destroys, rescuing critique tries to retrieve.'[36]

An important additional matter that Lewandowsky's analysis does not touch on is the relationship of Vergara's work to the development of what Marc Augé calls 'non-places'.[37] The Benjaminian image of the city as an accumulation of many layers of history ('wreckage upon wreckage') is, according to Augé, typical of a 'Baudelairean modernity', one that integrates earlier places into its landscape.[38] This image of the city relies on 'the presence of the past in a present that supersedes it but still lays claim to it'.[39] Augé quotes the following passage from Baudelaire's 'Tableaux Parisiens' as typifying this multifaceted image of the cityscape:

> [...] the workshop with its song and chatter;
> Chimneys and spires, those masts of the city,
> And the great skies making us dream of eternity[40]

The church spires and factory chimneys, Benjamin's arcades and Vergara's now-abandoned factories all co-exist in the temporally diverse fabric of a city. Contra this image, however, Augé's hypothesis is that 'supermodernity', his word for the most recent stage of modernity, produces 'non-places'. Unlike the places of Baudelairean modernity, these do not integrate earlier places.[41] Non-places – which include, for example, airports, supermarkets and highways – are constructed for clear purposes (transport, commerce, leisure), and they compel certain kinds of experience and behaviour.[42] In these purpose-built 'non-places' it is the text that directs individuals in their experience of the environment. For example, Augé writes about the well-designed *autoroutes* in France:

> [...] it is the texts planted along the wayside that tell us about the landscape and make its secret beauties explicit. Main roads no longer pass through towns, but lists of their notable features – and, indeed, a whole commentary – appear on big signboards nearby. In a sense the traveller is absolved of the need to stop or even look.[43]

This description is even more accurate of American highways. The visitor unfamiliar with the American landscape is faced with what appears to be a string of car parks, surrounded by b-rate identikit stores. These, one soon learns, are 'stripmalls', exemplary non-places. The relevance of Augé's claim that 'there is no room there for history unless it has been transformed into an element of spectacle, usually in allusive texts' becomes all too apparent when one passes by the large signs to 'historic sites' that adorn most American highways.[44] What reigns in non-places, Augé suggests, is only 'the urgency of the present moment'.[45] One might argue, instead, that what non-places efface is in fact the real urgency of the present moment, which can only be grasped in relation to what has been before and what might yet be. Non-places seem to exist in a kind of eternal present, where all one's needs (defined by the texts that direct both the needs and their satisfaction) are met, but where there is no sign of any before, or of any after.

To return to Vergara's photography, it is distinguished from the work of other photographers of urban ruin by his consistent use of a method known as retrophotography: he returns year after year (sometimes over a period of decades) to photograph the same site. He is thereby often able to tell compelling stories that might otherwise have been washed over and forgotten. A good example of this is found in a series of images from *The New American Ghetto* (1995), taken over sixteen years, during which a South Bronx housing complex was replaced by suburban-style townhouses.[46]

The final image (of eight images), taken in 1996, is of a row of identical, grey, picture-perfect suburban houses against a blue sky – a scene that Augé would no doubt classify as a characterless non-place. Looking at it, one can scarcely imagine that the landscape pictured in the first image (1980) – a six-storey, concrete block of apartments, most likely public housing – once existed in the same spot. It is not incidental that, while no people are pictured in the final image, the first in this series of shots from the South Bronx includes a group of residents (children, teenagers and adults) sitting, lounging, on the entrance stairway. There was life here, the images seem to say, whole communities connected to this milieu: do not forget us, we are not disposable. The images taken over the intervening years show the apartment block being demolished and replaced by the new, upmarket but soulless, housing.

What is most interesting about this series and others like it is that Vergara is doing more than witnessing forgotten layers of history embedded in the urban environment; he is witnessing the total effacement of these layers in supermodernity. It might be said that Vergara's work exists on the cusp of modern and postmodern landscapes. Where there are still layers of history to unearth in the urban fabric – where there is still essentially a 'Baudelairean modernity', in Augé's terms – Vergara engages in the Benjaminian project of rescuing and re-engaging the utopian energies that were contained therein. But where those layers of history have been entirely effaced, or are on their way to being effaced and replaced only with non-places, as in the series of images just analysed, Vergara's project might also be seen as bearing witness not just to particular histories but to the notion of history itself. Vergara's work, in these cases, is not just about documenting the hopes and dreams of various 'befores' but also about witnessing that there was a before, safeguarding even the very idea of a before.

A Jamesonian reading of Detroit 'abandoned spaces' photography

In existing discussions about 'abandoned spaces' photography in the popular press, both among reviewers of this photography and by the photographers themselves, there are often efforts to distinguish legitimate 'witness photography' from 'ruin porn'.

In the context of Detroit, the photographers who have come under the most sustained attack as proponents of 'ruin porn' are Marchand and Meffre, although many others have been heavily criticised as well. Journalist Andrew Sargus Klein writes about Marchand and Meffre's photo-essay 'Detroit in Ruins' (published in advance of their book) that

> it smacks of voyeurism and opportunistic journalism [...] poverty, systemic failures of government – and any sense of progress – all of this is overshadowed by the photogenic qualities of a good slant of sunlight cutting through the lobby of an abandoned public library.[47]

Reviewer Matthew Newton levels a similar criticism at Christian Burkert's photo-series 'Last Exit Detroit':

> Burkert does what hundreds, perhaps thousands, of photographers have done in recent years – he went on Safari in Detroit and turned his trip into a photo essay propagating the tale he wanted to hear. [...] What's problematic about this approach is that is does little but gawk at the cities

> and people in distress. [...] The glut of disaster porn photography currently cycling on the Internet has outsiders convinced Detroit is a post-apocalyptic wasteland, devoid of hope or humanity.[48]

By contrast, Vergara's work is often defended as more worthwhile, more critical. Both Lewandowsky and Leary, for example, defend Vergara's project above other ruin photography. Local Detroit photographers have also sometimes defended their own work as different in kind to photography by out-of-towners. An example of this kind of defence is found in J. Gordon Rodwan's *Detroit Is: An Essay in Photographs*. The accompanying essay by his son, John G. Rodwan, Jr, expresses distaste for ruin tourists who, 'in witnessing these emblems of misery, these monuments to failure, these icons of decay [...] expected both aesthetic and existential experiences'. However, Rodwan is keen at the same time to defend his father's photography from the wide-spread criticism levelled at imagery of Detroit ruin that it is nothing other than 'ruin porn': 'Just as not every picture of naked people counts as pornography, not every image of modern-day ruins arises from debased motives.'[49]

The core criticism most often levelled at Detroit ruin photography in these popular debates, that which photographers like Rodwan feel they need to defend their work against, is effectively the Jamesonian argument (this connection has not been made by previous critics). The claim at the heart of existing criticisms, in other words, is that this genre of photography employs the nostalgia mode to look longingly at a past (and indeed present) that its gaze in fact effaces. I will explore two key features of these Jameson-inflected criticisms – that the photographs lack context, and that they promote aestheticised passivity – and will argue that Vergara's work can escape the former criticism but not the latter.

That the vast majority of 'abandoned spaces' photographs are divorced from any real historical or present-day context is undoubtedly the case. Susan Sontag points out that 'photographs cannot themselves explain anything' and John Berger observes that 'the meaning of an image is changed according to what one sees immediately beside it or what comes immediately after it'.[50] In this regard, Vergara stands out. His consistent effort in documentation and attention to the specific context of each image is notable. As Leary argues, Vergara

> addresses the ahistorical failing of much ruin photography by investing much more heavily in the ruins: Revisiting the same site over a period of years, or even decades, Vergara's pictures show with often heartbreaking clarity the slow, painful transformation of a house, a street, a neighborhood.[51]

However, this level of detailed context is, to varying degrees, lacking from subsequent Detroit photographers.

To demonstrate this, it is worth comparing four photographic presentations of the infamous Packard Automobile Plant in Detroit. In Vergara's book, *American Ruins*, images of the old factory appear in a section clearly titled 'Packard

Automobile Plant, Detroit'. The six pages of images in this section include close-ups and long shots, from various angles, and were taken in 1991, 1993 and 1998. The accompanying text includes a history of the plant (mentioning that it was designed by the distinguished architect Albert Khan in 1903, that in World War II it produced Rolls-Royce airplane engines, and that it stopped producing automobiles in 1956); a description of the now-abandoned plant and of the illegal 'rave' parties that were often thrown inside it; and an interview with the property manager who had responded angrily when Vergara called Packard a ruin. The inclusion of these interviews is characteristic of Vergara's attempts to record the complexity of feelings among local residents towards the places he photographs and towards his project.[52]

By contrast, Hank Van Rensbergen's *Abandoned Places: The Photographer's Selection* (2010) also devotes four pages to the Packard Plant, Detroit. Each image is a long shot of the interior of the plant: crumbling windows visible at the side, debris on the ground and – in two of the shots – abandoned, graffiti-covered vehicles as the ostensible 'subject'. The Packard Plant images are accompanied by no text, not even a subject line: it is only by looking to the 'photo index' at the back of the book that one is able to ascertain where the images were shot. They are indistinguishable from images of other abandoned locales.

Tong Lam's *Abandoned Futures: A Journey to the Posthuman World* (2015) goes one step further: the very beautifully shot photographs have no subject lines, and there is no photo index at the back. Tellingly, the book does not even have page numbers. There are several images which might be of the Packard Plant, particularly given that the accompanying essay does at one point mention the plant by name as 'the largest abandoned factory in the world', but other industrial ruins are also mentioned and it is difficult to be certain which the image is of. Lam's book can be defended against the criticism that these 'abandoned spaces' photography projects lack historical context on the ground that, unlike other photographers of this kind, his aim is not documentation or witnessing. The accompanying essay in his book is a thoughtful and philosophical reflection on 'abandoned spaces' imagery in which he explores the fascination with these images. He is fully aware of the criticisms levelled at this genre of photography:

> When modern ruins are divorced from the troubling contexts in which they are produced, they evoke an abstract notion of beauty and nostalgia. Some critics have therefore referred to this genre of photography as 'ruin porn', meaning that its audience tends to passively consume these images as mere spectacle. These critics may be right, but only to a degree. After all, is it really possible to appreciate these images without thinking about the condition of our own existence?[53]

That Lam seems more interested in posing philosophical questions about ruination in the modern world than pursuing a project of historical witnessing is clear from the poetic chapter titles he chooses, which include

'Time Speeds up', 'Industrial Sublime', 'Maximizing Life' and 'Post-Apocalypse Now'.[54]

But the net impact of the tendency towards de-contextualisation of Detroit 'abandoned spaces' photography is that the images convey pseudo-historical depth without containing any actual historical content: a clear example of Jameson's 'nostalgia mode'. Leary puts this criticism well:

> Taken together, all the images of the ruined city become fragments of stories told so often that they are at the same time instantly familiar and utterly vague, like a dimly remembered episode from childhood or a vivid dream whose storyline we can't quite remember in the morning: Murder city! Unemployment! Drugs! White flight![55]

Leary argues that Detroit 'abandoned spaces' images become 'pictures of historical oblivion' and function as 'pictures of nothing and no place in particular'.[56] This rings especially true when the Detroit images are presented, as is often the case, alongside images from, for example, old sanatoriums in ex-soviet countries or Nara Dreamland in Japan: with little or no explanation, these distinct histories are all thereby reduced to the same fate. Susan Sontag's description of the potential of the photograph to present 'a view of the world which denies interconnectedness, continuity, but which confers on each moment a character of mystery' seems somehow to reach an apex in Detroit 'abandoned spaces' images, where the very real fallout from the pursuit of economic growth, and stories of systematic neglect, become alluring, romantic, mysterious, beautiful – and utterly banalised.[57]

Writing of Stanley Kubrick's *Barry Lyndon*, which purports to be a historical film, Jameson remarks that its

> very perfection as a pastiche intensifies our nagging doubts as to the gratuitous nature of the whole enterprise. Why this 18th Century at all in the midst of a late 20th Century culture? Or, in that case, would almost anything else not have done just as well (a Kubrick Elizabethan era, a Kubrick American Revolution, a Kubrick *Ivanhoe*)? […] And as for the choice of subject, why should a Southern small town, a California university, or the Manhattan of the 1970s be any less arbitrary a starting point, in a fragmented multi-national culture, than the London or German principalities of this 18th Century?[58]

The same nagging doubt is precisely what perturbs many viewers of contemporary 'abandoned spaces' photography. As Leary says:

> One often finds oneself asking of *Detroit Disassembled, The Ruins of Detroit*, and indeed all ruin photographs, first, 'What happened?' followed swiftly by, 'What's your point'?[59]

In my Benjaminian reading I suggested that Vergara's work is caught between the photography of modern landscapes (where his aim is to rescue latent histories and forgotten possibilities from the layered fabric of the city) and of postmodern landscapes (where he undertakes the even more urgent task of documenting the mere fact that there was ever a 'before' of the non-places of supermodernity). Compared with Vergara's project, are these more recent 'abandoned spaces' photography books too immersed in the culture they attempt to critique? Photographers like Hank Van Rensbergen and Matthew Christopher might think, like Vergara, that they are witnessing or documenting a 'before', but more often than not the only thing to which their books testify is the ease with which historical specificity is obscured by a beautiful slant of sunlight.

A final point about context concerns the ways in which people are presented in the images and by the accompanying texts. James Griffioen, a Detroit-based photographer and blogger (www.sweet-juniper.com) often credited as the father of the term ruin porn, differentiates his own photographic practice from that of others: 'I take pictures of ruins, too, but I put them in the context of living in the city.'[60] His criterion, 'living in the city', is understandable but arbitrary: surely some photographers living in Detroit might be labelled exploitative, while others who live elsewhere, like Vergara, might demonstrate a real commitment to the locale. We would do better to ask whether a given photographic practice strives to place or frame ruins in the context of a living city.

Vergara's work clearly does this. Not only is his text full of references to current residents and the impact of dereliction on them, but many of his images also include local residents, thus testifying visually to the realities of life lived among these urban ruins. The first image in the introduction to *American Ruins* (1999), for example, of an old gas station in Gary, Indiana, pictures ten or so children, African-American boys, playing basketball at the site of ruin – a testament to life continuing in these locales, which sets the tone for the rest of the book.

A useful contrast is offered by Hank Van Rensbergen's *Abandoned Places: The Photographer's Selection* (2014). There is not a single person in any of the images in the entire book. In his defence, Van Rensbergen's primary interest is in urban exploration, so many of the places he photographed may have been difficult to access, thereby precluding local residents from entering. But the character of his text is also telling. The following passage occurs in his short description of Detroit:

> There is something strange about this city. There are hardly any cars driving on the wide, six lane roads. Three cars are parked in the car park and the woman by the barrier is staring absently into space. On the pavement, a homeless person is wandering around aimlessly.
>
> The other skyscrapers around me stare at me with hollow eyes: most of the buildings are empty above the first floor, and you can sometimes see a curtain flapping out through an open window.[61]

There is little distinction here in the way that Van Rensbergen describes human beings and buildings. Both people and skyscrapers are absent, aimless, staring into space, with hollow eyes. The buildings are humanised and the people dehumanised. There is no sense that the way a woman stares is not, in fact, a reflection of or akin to the way a building stares. Van Rensbergen projects onto the unnamed woman and homeless person the story he wants his images to tell (one of spooky, eerie abandonment), just as he projects onto the building a pair of 'hollow eyes'.

This dehumanisation of the people in Detroit is all too familiar. Julian Temple, the director of the documentary *Requiem for Detroit*, describes his experience of the city for an essay in the *Guardian*. In the shadows of the derelict skyscrapers, he writes, 'the glazed eyes of the street zombies slide into view, stumbling in front of the car'.[62] The sentence would not be out of place as the shot description in a post-apocalyptic screenplay, except of course these are in fact real people, not zombies.

Just as the people are dehumanised, objects in Van Rensbergen's images, as well as his text, are humanised in order to conjure up the mood he wishes to evoke. His photographs often foreground a solitary object – a chair, car, mask or piano – that seems to plead of loss and abandonment. Masks and vehicles feature prominently: the former are ready-made to invite personification, while the latter are often framed so that the headlights seem to stare with precisely those 'hollow eyes' Van Rensbergen sees in his skyscrapers.

This technique of personifying objects to evoke the desired mood, rather than photographing actual people (local residents, homeless people, ravers) to record the actual mood, is common among 'abandoned spaces' photographers. Tong Lam, for example, gives an empty office chair centre stage in one photo of an abandoned locale, encouraging the viewer to imagine the ghostly presence of a long-gone figure. In another, an upturned carrier of some kind (presumably one used to move things from place to place) is corpse-like, bringing to mind the husk of a dead beetle lying on its back.

Photographs can be used to evoke all kinds of feelings and there is no absolute moral or aesthetic obligation for photographers to aim at historical accuracy or depth. However, the use of personified objects to tell the story of nostalgic loss and beautiful abandonment is highly problematic if, as often happens, the photographs are claimed to constitute some kind of historical testimony. Even when this is not the case, as in Tong Lam's book, the collective impact of these repeated visual tropes is to blur distinctions between the intentions of the various projects, thereby ultimately affecting the genre as a whole, which leads me to the second line of attack levelled at these photographers: aestheticised passivity.

This is perhaps the more damaging criticism, since even Vergara's work cannot escape it. There are two key elements to this charge, both of which are central to Jameson's critique of the nostalgia mode (and which are voiced most clearly in his essay on *The Shining*): the images are beautiful and they are boring.

The beauty of the images is hard to deny; even the harshest critics of the genre, both journalists and academics, admit as much. For example, journalist Sargus Klein writes about the Marchand and Meffre photo-essay that 'with the slimmest of contexts, the viewer is left to simply enjoy the sixteen *beautiful* shots of the Grotesque, the Terrible Decline'.[63] Dora Apel's book, an attempt to understand the fascination with Detroit ruin photography, is titled *Beautiful Terrible Ruins: Detroit and the Anxiety of Decline*. As Apel notes in her book, the so-called pornographers, the out-of-towners most heavily criticised for their Detroit photography, also often have better equipment than the locals, thus enabling them to render their images more beautiful.[64]

The beauty of the works in nostalgia mode is something to which Jameson is highly alert, since this quality is central to his understanding of the 'cult of the glossy image', as discussed in Chapter 5. Jameson is conscious that beauty can be dangerously addictive, can lose its charge and lead to a kind of repetitive boredom, as the following description of Kubrick's films suggests:

> Think now [...] of the 'beautiful' in Kubrick's work: one still obsessively recalls the sound of 'The Beautiful Blue Danube' that spins the slowly rotating space shuttle in *2001* on its way to the Moon, like a Musak in a high-class elevator soothing and tranquilizing both the official bureaucratic travellers on the vehicle and ourselves as well [...]. The high-cultural banality of the waltz thus expresses the banality of this harmonious UN-run global world as well as the boredom of its depthless inhabitants.[65]

But, as well as being beautiful, 'abandoned spaces' images are by now so ubiquitous that they have also become boring. This is evidenced by the kinds of commentary on the phenomenon that appear in art magazines, blogs and news articles. John Cunningham, for example, writes that: 'In the image-world of hopefully "late" capitalism the industrial ruin has acquired a fair amount of cultural capital and such spectacular over-determination is a major reason for ennui with corroded concrete.'[66] He even cites a poem that uses an automatic text generator to describe his boredom with urban decay, giving voice to this widespread experience of repetition and boredom:

> Sick of ruins/sick of meaning ruins/ruined/decay/blight/derelict/poetry/heavy bricks/getting heavy/sick of work/getting sick/labour history/dead city/history dead/city labour/dead city/invading ruins/my apologies/my theft/sick of poverty/sick of ruins ...[67]

Detroit photographers often take pictures of the same abandoned buildings, even the same rooms, sometimes from the same angles. The overall impact of seeing these images again and again on different publication platforms (on news sites, personal websites, Tumblrs, Instagram, Facebook, and in coffee-table publications and photography shows) is of overwhelming repetitiveness.

Michigan Central Station is perhaps the best-known location photographed by photographers of abandonment. In the four coffee table books under discussion, only Vergara's image of the station (in *American Ruins*) stands out, in that in it is presented alongside multiple interior and exterior shots of the station, and is accompanied by a lengthy text description as well as additional photographs of people such as 'Catfish', a three-year resident of Michigan Central. But precisely this context is lost when images are re-shared or re-created on social media and other 'abandoned spaces' web pages, as they so often are.

More striking than the repeated use of the same locations, is the similarity of images of different locations. Long shots of abandoned factory rooms, broken windows running along the side, rubble on the floor: these are staples in nearly every book of this kind. Moreover, image-saving applications like Pinterest have thousands of such images tagged under 'abandoned factories', re-posted time and again. The origin and subject of these pictures is usually unclear, but a quick search of Pinterest gives an idea of the scale on which similar imagery is being produced and shared, and the numbing effect that this has on the viewer.

'Abandoned spaces' photographs also often deploy the same visual tropes. The clearly visible slant of sunshine in an otherwise darkened room is a favourite. Other common visual tropes include the lone chair left in an otherwise empty room; the abandoned cinema or theatre; the staircase; leaves scattered or foliage growing in or around the building; discarded teddy bears and dolls; and former hospitals, particularly psychiatric hospitals, which add an eerie affect to the images. Another, more baffling, common trope is a mask lying on a floor or table. As suggested in the analysis above, this serves the purpose of humanising the spaces to tell a particular story of abandonment and loss, but it is not clear why such masks have so frequently been left behind on the floors of abandoned buildings. This is not to suggest that they have been planted, only that, however uncommon a feature of modern ruins they in fact are, they provide a visual metaphor that has been frequently deployed by photographers of these abandoned buildings. Each of these symbols could be analysed further to understand their meaning or tone, but the point here is simply to demonstrate the extent to which the photographs mimic one another. This is important, because images that were meant to be arresting lose their power when the viewer is desensitised by numbing repetition. In this regard, the term ruin porn, which has become so prevalent in describing 'abandoned spaces' photography, seems accurate; the endlessly repeated images have much the same numbing effect as pornographic images. Furthermore, this widespread repetition affects the impact and reception of all abandoned spaces photography, regardless of the original context or intent of the photographer.

Crucially, this 'aestheticised passivity' undercuts some of the power of a Benjaminian reading of Detroit 'abandoned spaces' photography. Both Edensor's and Armstrong's accounts of their own encounters with actual abandonment (on a Manchester commute, or in small-town Canada) rely on Benjamin's notion of shock – the belief that 'when the past is reflected in the dewy fresh "instant," a painful shock of rejuvenation pulls it together once more'.[68] To recall the past as it is always-already-known cannot awaken us; the only moments that have the potential to do this are those which 'seize hold of a memory as it flashes up at a moment of danger'.[69] This is why Armstrong emphasises the importance of unplanned wandering. Edensor, too, sees the value of the absent-present-pasts that hide within the mundane familiar landscape in their power to '*erupt* into the present' and to disrupt 'everyday routines, habits and unreflexive practices'.[70]

Most Benjaminian accounts of the critical potential of abandoned spaces rely at least in part on the capacity of the past to shock and disrupt, but it is harder to claim such a capacity for Detroit's 'abandoned spaces' photography in the present cultural context. David Gross argues that the dominant type of social memory today is the memory of the ephemera of popular culture, which is fostered and sustained by mass media and which enables people to remember in a conformist and unchallenging way.[71] While he encourages acts of remembering divergently, seeking out counter-memories, it is questionable how easy such acts are in an age in which so much is already assimilated through the virtual circulation of memory. Certainly 'abandoned spaces' photographs are by now no more jarring on a Pinterest board than pictures of the latest wedding fashions. 'Abandoned spaces' photography has, in short, become too fashionable, the aesthetic too much in vogue. These once powerful, divergent images have been incorporated in present-day consumer logics.

This is a problem that gets at the heart of a Benjaminian reading of contemporary nostalgia texts. Mark Fisher puts it starkly:

> This dyschronia, this temporal disjuncture, ought to feel uncanny, yet the predominance of what Reynolds calls 'retromania' means that it has lost any *unheimlich* charge: anachronism is now taken for granted.[72]

Benjamin was himself acutely aware of the capacity of the mind to quickly incorporate initially shocking or divergent impressions into existing mental frameworks, as he alerts us:

> The greater the share of the shock factor in particular impressions, the more constantly consciousness has to be alert as a screen against stimuli; the more efficiently it does so, the less do these impressions enter experience (*Erfahrung*), tending to remain in the sphere of a certain hour of one's life (*Erlebnis*).[73]

There is still merit to a Benjaminian reading of Vergara as having performed some kind of rescuing critique, but the additional possibility that 'abandoned spaces' images have the power to disrupt presently dominant logics is lessened by the sheer ubiquity of these images, a ubiquity which doubtless reduces their anachronistic charge.

Several of Van Rensbergen's photographs are of an abandoned hotel, the Overlook Hotel, in Germany.[74] That it has the same name as Kubrick's

Overlook Hotel in *The Shining* is fitting. Just as Jameson, in his analysis of *The Shining*, describes this once grand hotel as having become 'as multinational and standardized as a bedroom community or a motel chain', so 'abandoned spaces' imagery is increasingly standardised.[75] Once-powerful photographs of imposing structures are now indistinguishable in content and feel from one another. Indeed, a quick Google search for 'abandoned spaces photography' certainly seems to lend credence to a Jamesonian view of postmodern culture as a culture of 'surface, simulation, fragmentation and instantaneity'.[76]

A Blochian reading of Detroit 'abandoned spaces' photography

If a Benjaminian reading struggles to overcome the problem of aestheticised passivity, I believe that a Blochian reading can offer a different way of approaching the photographs: one that does not rely as directly on the capacity of 'abandoned spaces' to shock and thereby disrupt present-day logics. Thus far, I have focused mainly on what photographs of Detroit's 'abandoned spaces' achieve. Do they witness history or efface it? Do they shock – jolting the viewer into new recognitions – or do they numb? But there is another set of arguably even more important questions to ask: Why are these images so appealing to contemporary viewers? Why have they become increasingly popular over the last two decades? Why do they fascinate?

Several journalistic commentators have suggested answers to these questions. Kyle Chayka argues that Detroit ruin porn functions as a 'memento mori', 'a call to remember that the same fate as Motor City could befall all our great cities, all of unstable accomplishments'. Looking at ruin porn is, in his view, akin to 'gawking at the possibility and the danger of death'.[77] Eli Rosenberg also asks: 'Is Detroit the car crash we can't take our eyes off? Is it wrong of us to stare?'[78]

In academic work on the subject, too, the question is increasingly posed. Dora Apel argues that the fascination with Detroit ruin photography is a manifestation of latent fears about cultural decline. She claims that 'the anxiety of decline may be understood as the dark side of modernity, which is founded on a set of universalist values stemming from the Enlightenment that supported ideals of progress and rationality through science and technology' and that Detroit has become a central locus for this anxiety of decline.[79] The images, Apel argues, work to assuage our anxieties by aestheticising ruin, thus allowing us to cope with our fears: 'beauty making the terrible bearable'.[80] Apel's explanation is different to Chayka's in that it is more concerned with mastering anxiety than with titillation, but both take as a premise that fascination with Detroit's 'abandoned spaces' arises out of a fear that what has happened to Detroit might happen to us.

Another view is found in Steinmetz's argument that the lure of Detroit's 'abandoned spaces' has to do with a more straightforward nostalgia for the Ford era. Buildings like Michigan Central Station do seem to project a certain confidence and vision of a prosperous society, and it is not difficult to see how one might read nostalgia for this prosperous past into our fascination with Detroit's ruins. This theory is similar to Jameson's reading of *The Exorcist* (1973) as expressing 'nostalgia for a system in which Good and Evil are absolute black-and-white categories'.[81] Interestingly, this theory of the allure of 'abandoned spaces' is directly counter to a Benjaminian reading: rather than seeing Detroit's 'abandoned spaces' as having a temporally divergent charge that poses a critical challenge to oversimplified notions of progress, the position here is that the once-great buildings of Detroit are nostalgically reminiscent of an uncomplicated belief in progress that is no longer easy to inhabit.

All of these theories seem viable, and may on some level be valid. As DeSilvey and Edensor remind us, 'ruin meanings are continually up for negotiation'.[82] Importantly, as Bloch's analysis of the non-contemporaneous longings of Weimar Germany shows us, a longing for or fascination with some kind of past can contain multiple and conflicting desires. The work of the critic, as he sees it, is not to reduce works (or responses to those works) to one definitive thing or another, but rather to mine them for that which is hopeful: to seek out the gold in the rubble.

Bloch's approach, as already discussed, is a useful counter to Jameson's critique of the nostalgia mode (which effectively denounced it altogether as ideology) since it allows for a more nuanced understanding of this mode as a complex mix of sometimes contradictory longings: containing, in other words, both ideological and utopian content. But Bloch's approach is also useful in addressing the problem facing Benjaminian interpretations of nostalgia that rely on the idea of a Proustian shock of recognition: that the temporally divergent (non-contemporary) has become so familiar to present-day subjects that it has lost its anachronistic charge. In mining the longings of 'secret Germany' Bloch does not rely as heavily as Benjamin on the potential of past material to shock and awaken us from a life lived in automaticity. Bloch mined the nostalgic ideology for latent desires, still-not-yet-fulfilled utopian possibilities, 'a muffled non-desire for the Now'.[83] That these utopian aspects are muffled, hidden within ideology, rather than jarring suddenly with our habituated perspectives, does not mean that they are any less potentially powerful. It does, however, make the work of the critic pressing. These desires, which do not immediately shock or awaken but rather hum along quietly – a faint tune that is discordant with the noisier song of contemporaneity and can only be heard with an attuned and attentive ear – can be appealed to only if they are heard and understood.

If Benjamin's search for the 'hole in the whole' led him to a fascination with the anachronistic, the marginal and the forgotten, Bloch was somewhat more prepared to mine even seemingly mainstream cultural practices and products for hope-content that was yet to be sifted out. So when the anachronistic is robbed of its anachronism and the marginal is made mainstream, as is the case with Detroit 'abandoned spaces' photography, a Blochian approach is particularly helpful.

Returning to our present subject, what this means is that even if the potential shock-factor of these images is diminished by incessant repetition, and they are consumed in 'aesthetic passivity' (which is undoubtedly the case), that does not mean that they do not contain submerged hope-content. There are two distinct strains of latent utopian content in these images and the fascination with them to which I want to draw attention: a resistance to utilitarian consumer logic, and a thwarted desire for historical depth.

The symbolic allure of abandonment

The problem with theorists who read the popularity of Detroit 'abandoned spaces' photography as speaking to particular fears about American decline, or as a longing for a particular moment of Fordian prosperity, is that they ignore the very abstract context in which this imagery is so often consumed. As already mentioned, photographs of Detroit's abandoned factories are consumed alongside photographs of, for example, ex-Soviet sanatoriums or abandoned fairgrounds in Japan. These images reflect very different histories and stories of abandonment – the failed dream of the communist era is hardly the same as the failed promise of the Detroit car manufacturing industry – and yet they appear to be motivated by an indistinguishable romanticism or longing. Although the interchangeability of these images is rightly viewed as problematic insofar as the images obscure the specific histories in question, we might still ask what all of these images have in common: Why do they appeal to viewers in the same way? Against readings which focus on the tale of American decline, I propose reading Detroit 'abandoned spaces' imagery as part of the broader neo-romantic nostalgia mode that has been discussed in different guises throughout this book. Only by abstracting from the particulars, in this case, does the broader contemporary allure of abandonment make itself clear. What do all of these images of 'abandoned spaces' have in common? Simply, but importantly: they picture that which is no longer useful. And this, I think, is key.

Vergara's work might be understood not only as the work of historical witnessing, but also as a different kind of rescuing: the salvaging of the no-longer-necessary. Indeed, the meticulous documentation that Vergara undertakes is in a way reminiscent of Tavi Gevinson's archiving of her own memories and the details of her suburban existence, discussed in Chapter 4. Vergara tries to be objective, whereas Tavi's systematisation is shaped only by her mind's contours, but both attempt to salvage things regardless of their utility. For Tavi, things which might otherwise be seen *only* as objects of use (a hairpin, a carpark, the road to school) are re-configured according to her own 'strange magic' and thereby in some sense saved; for Vergara, things that are no longer useful and are thus en route to being obliterated, are somehow saved from oblivion. Of course, they are not actually saved in any historical, material sense, and therein lies the key criticism against ruin photography: the poverty persists. But the photographs of abandoned places confront those places deemed useless and in finding beauty there they express a utopian resistance to the relentlessly utilitarian logic that characterises contemporary capitalism. The photographs capture the allure of what would otherwise be discarded as the refuse of history, and they bring the viewer into an encounter with the useless (an encounter all too rare in a world governed by consumer logics) that itself functions in some way as a cry in the dark against the contemporary emphasis on utility. This is something that can be claimed also of photographers of abandonment subsequent to Vergara (even if his success as a historical witness can not).

Žižek ponders a similar idea in his film *The Pervert's Guide to Ideology*. At one point, he visits the world's largest dumping ground for abandoned aeroplanes, deep in the Mojave desert. First, he admits the obvious: this dumping ground is a visceral testament to the tremendous amount of waste produced in our current global system. But then he asks whether this waste doesn't nonetheless serve an important symbolic purpose: it highlights that there are things in the world that serve no discernible end, and this allows a momentary suspension of the logics within which we otherwise exist. He muses:

> What moments like confronting the planes in the Mojave desert bring us is maybe the chance for an authentic passive experience? Maybe without this properly artistic moment of authentic passivity nothing new can emerge? Maybe something new only emerges through the failure, the suspension of proper functioning of the existing network of our lifeworld?[84]

Abandonment is only romanticised by people who do not suffer its realities, and there is an understandable disdain for the out-of-towners who come to photograph the broken parts of Detroit in which the largely African-American, largely poor, citizens live (Detroit is the city in the United States with the largest proportion of African-American citizens at 82.7 percent).[85] But in a Blochian reading it is possible both to criticise 'abandoned spaces' photography for naturalising and aestheticising what is in fact the brutal fallout of the pursuit of profit and, at the same time, to ask what legitimate hopes might explain the allure of these images. And although it is undeniable that abandonment is seldom romanticised by those who must live in its wreckage, it is also the case that abandonment can only seem appealing in a culture where space is so seldom left alone, where it is so seldom un-utilised. The only places that do not seem to be open to co-option in pursuit of profit are places that have already been co-opted and used up, and perhaps it is this which explains, at least in part, the huge symbolic allure of 'abandoned spaces': they are in some sense 'a space outside'.

Nostalgia for the vertical

I want to return to Marc Augé's distinction between the landscapes of Baudelairean modernity and the landscapes of supermodernity. In the former, layers of history accrue; church spires, factory chimneys and skyscrapers co-exist in the multifarious fabric of the city. In the latter, purpose-built non-places (airports, highways, supermarkets, suburban housing complexes, chain restaurants) efface everything for which there is no longer a use. These non-places of supermodernity are undifferentiated. One does not stumble through or upon them, and there are no traces of outmoded layers on which one might unexpectedly alight; one moves through them as directed to do so by the text that signals the intended utility of the space. One is reminded also in thinking about the 'non-place' of Jameson's comment about the Westin Bonaventure building in Los Angeles, an exemplary non-place, that 'it does not wish to be part of the city but rather its equivalent and replacement or substitute'.[86]

J. M. Coetzee, in an essay about the picturesque, argues that in European art of the nineteenth century 'the sublime is far more often associated with the vertical than the horizontal', and that sublime art 'makes verticality – heights and depths – the locus of important – profoundly important – feelings such as fear and ecstasy, and values such as transcendence and unattainability'.[87] Coetzee is talking about natural vistas – mountains or plains – in landscape paintings; but I think his distinction between the horizontal and the vertical is also important in relation to contemporary urban and suburban landscapes, and can help to illuminate the genre of 'abandoned spaces' photography that is my concern.

Jean Baudrillard's *America*, a 1984 book, details his own experiences in and reflections on 'America' as symbolised in its cultural imaginary, and as seen from the perspective of a European. In it, he often suggests that the truly 'American' landscapes are horizontal. In particular, he claims that a certain kind of 'America' finds its aesthetic form in the expansive deserts of the American West:

> I went in search of *astral* America, not social and cultural America, but the America of the empty, absolute freedom of the freeways, not the deep America of mores and mentalities, but the America of desert speed, of motels and mineral surfaces.[88]

I believe that the America that Baudrillard means by '*astral* America', the America 'of the empty', of 'freeways', 'motels' and 'surfaces', is the landscape of what Marc Augé, a decade later, termed non-place. Baudrillard finds the ultimate aesthetic approximation of *astral* America, the landscape of 'non-place', in the desert. It is important to emphasise that Baudrillard's writing does not imply that the actual desert is a 'non-place' in Augé's sense of the term; clearly it is not (the desert is not purpose-built or written over by signs, and it serves no immediate utilitarian function). But the affective experience that Augé would later associate with 'non-place' (the landscape of highways, airports and motels) certainly approximates Baudrillard's affective experience of the desert's eternal sameness. Baudrillard comments:

> The silence of the desert is a visual thing, too. A product of the gaze that stares out and finds nothing to reflect it. There can be no silence up in the mountains, since their very contours roar. And for there to be silence, time itself has to attain a sort of horizontality; there has to be no echo of time in the future [...]. Desert: luminous, fossilized network of an inhuman intelligence, of a radical indifference.[89]

The absolute horizontality that Baudrillard describes here is a reflection of the absence of an 'echo of time in the future', and it is this absence that is the defining feature of non-places (which is why the desert as non-place is such an apt metaphor). Non-places know only the present moment: a present absent the past and accordingly absent any sense of a future. The idea of 'radical indifference' is important, too. The 'gaze that stares out and finds nothing to reflect it' is a powerful description of the absence in non-places both of an historical other against which to push (signs of a before, distinct from today, that might lead to another tomorrow), and of the signs of distinctive or unique human labour, agency and creativity.[90]

Baudrillard explicitly contrasts the absolute horizontality of American civilisation with European civilisation, but we should be wary of taking this position at face value. It is true that, as Coetzee points out, European art has historically located the sublime in the vertical, the mountainous – where the very contours roar. But early American landscape painting such as that of the Hudson River School did the same, and in any case the real distinction that Baudrillard is making is not between 'Europe' and 'America' as essentialised categories (although he does arguably at times fall prey to this) but, more poignantly, between the 'Old World' and the New World'. He writes:

> The form that dominates the American West, and doubtless all of American culture, is a seismic form: a fractal, interstitial culture, born of a rift with the Old World [...]. The only question in this journey is: how far can we go in the extermination of meaning, how far can we go in the non-referential desert [...]?[91]

Non-place is not a uniquely American phenomenon. The highways that Augé describes are French. Highways, airports, suburbs, motels, malls – these things are not all American inventions, but they do arguably reach their apex in America. Americanisation is of course, on one level, synonymous with globalisation ('In the very heartland of wealth and liberation you always hear the same question: "What are you doing after the orgy?" [...] This is America's problem and, through America, it has become the whole world's problem').[92]

But even more than a distinction between the Old World and the New World, or between the local and the global, the distinction between the vertical and horizontal is at heart a distinction between place (a landscape of depth) and non-place (a landscape of surface and simulation). When Baudrillard writes about New York that it can only muster verticality as a kind of 'special effect', the real point is not to emphasise its American-ness, but its relative new-ness:

> Every special effect can be found here, from sublime verticality to decay on the ground [...]. In years to come cities will stretch out horizontally and will be non-urban (Los Angeles). After that, they will bury themselves in the ground and will no longer even have names. Everything will become infrastructure bathed in artificial light and energy. The brilliant superstructure, the crazy verticality will have disappeared. New York is the final fling of this baroque verticality, this centrifugal excentricity [*sic*], before the horizontal dismantling arrives.[93]

Here New York becomes the last outpost of a verticality – if only as 'special effect' – that is essential to Baudelaire's Paris or Benjamin's Berlin. But even this verticality is threatened by a drive for expansion that can only ever entail a levelling out; or at least this is Baudrillard's suspicion.

The contrast between the verticality of Baudelairean modern places and the sheer horizontality of supermodern non-places is well captured by comparing two of director Wim Wenders' films: *Wings of Desire* (1987) and *Paris, Texas* (1984).[94] In *Wings of Desire* (1987) the camera pans up and down over the heights and depths of the city, Berlin. The angels sit atop statues and church spires, at varying levels of the many library floors, or at times underground in the city's tunnels. Like Benjamin's 'Angelus Novus' they can look down at or up through the layers of history, piled wreckage upon wreckage, that are embedded in the fabric of this city, a city Benjamin wrote about in depth. Indeed, the German title of the film, *Der Himmel über Berlin* (The Sky over Berlin) captures this sense of looking down from heights to depths. By contrast, the non-places of supermodernity – the car parks, motels and highways – are represented in Wim Wenders' *Paris, Texas* (1984) by sheer horizontality.[95] The camera pans sideways along endless stretched landscapes, uninterrupted in its reach (the 'non-referential desert').

The film stills from *Wings of Desire* and *Paris, Texas* shown in Figure 6.1 demonstrate something of the difference between these perspectives. The implied location of the camera is especially interesting: in *Wings of Desire*, it seems often to be suspended in mid-air, hovering above or below the focal point, reproducing the movement of the angels themselves; in the latter, the camera has its feet on the ground, so to speak. The gaze of the characters, too, is not up or down but straight ahead or sidelong.

To return to my primary focus, I propose that the fascination with 'abandoned spaces' imagery has to do in part with a nostalgia for the vertical: a longing, that is, for historical depth in a culture which threatens to efface (literally, build over) the past (the outmoded) altogether. The slanted rays of sunlight which are so frequently captured sliding into these forgotten spaces are suggestive of a sublime verticality: they cut diagonally across the images,

Figure 6.1 Wim Wenders film stills: verticality and horizontality.

drawing viewers' attention both to the height of these often imposing structures and to the forgotten nooks and crannies. Staircases, too, are frequently depicted in these images, suggestive of yet more layers and unseen levels above and below. By contrast, the eternal horizontality of Baudrillard's desert or Wenders' *Paris, Texas* has no need for staircases: there is no up or down, no yesterday or tomorrow, only today stretching out interminably. Moreover, the very fact that abandoned places still exist recalls the possibility of a more historically layered landscape, of places which still have traces, as opposed to non-places, where all trace of a 'before' has been erased. The longing elicited by these photographs of formerly glorious train stations and vacated factories is on some level a nostalgia for a Baudelairean modernity and its layered cityscapes.

So although, with the arguable exception of Vergara, most photography in this genre fails to grapple seriously with historical time, it nonetheless expresses a muffled longing for the possibility of real history: of a tomorrow distinct from today and yesterday. This longing is expressed as a nostalgia for the vertical, for the layered cityscapes of Benjamin and Baudelaire. It could be said that this genre of photography, although mostly operating within the nostalgia 'mode', still exhibits a kind of nostalgia for the nostalgia 'mood'. The nostalgia for the vertical is a nostalgia for the deep tussles with time, history, loss and future possibility that Benjamin was able to salvage from the rubble of the city, but that recedes further and further from view, like a mirage in the distance of the 'deserts of meaninglessness'.[96] In this nostalgia for the vertical, this muffled non-desire for the Now, there is, however faint, the refusal to give up on that which is harder than ever to access: the real sense of history, with its awareness of a tomorrow distinct from today and a capacity to push against the particular constellations of yesterday.

Notes

1 Jeffrey Eugenides, 'The Art of Fiction no. 215, Interview by James Gibbon', *The Paris Review*, 199 (2011) www.theparisreview.org/interviews/6117/the-art-of-fiction-no-215-jeffrey-eugenides [accessed 4 September 2014].
2 Tim Edensor and Caitlin DeSilvey, 'Reckoning with Ruins', *Progress in Human Geography*, 37.4 (2013), 465–485 (p. 465).
3 Andreas Huyssen, 'Nostalgia for Ruins', *Grey Room* (2006), 6–21 (pp. 7–8).
4 Tim Strangleman, '"Smokestack Nostalgia", "Ruin Porn" or Working-Class Obituary: The Role and Meaning of De-industrial Representation', *International Labor and Working-Class History*, 84 (2013), 23–37.
5 Julia Hell and Andreas Schönle, *Ruins of Modernity* (Durham, NC: Duke University Press, 2010), p. 4.
6 John Patrick Leary, 'Detroitism', *Guernica*, 15 January 2011, www.guernicamag.com/features/leary_1_15_11/ [accessed 3 June 2016].
7 Jamie Rann, 'Beauty and the East: Is it Time to Kick Our Addiction to Ruin Porn?', *The Calvert Journal*, 31 July 2014, http://calvertjournal.com/features/show/6168/power-and-architecture-ruin-porn-photography-rebecca-bathory [accessed 3 June 2016].
8 Camilo José Vergara, *American Ruins* (New York: Monacelli Press, 1999).
9 Yves Marchand and Romain Meffre, *The Ruins of Detroit* (London: Steidl, 2010).
10 Henk Van Rensbergen, *Abandoned Places: The Photographer's Selection* (Tielt: Lannoo, 2014). It should be noted that there are several other photography books with the same name, for example *Abandoned Places* (2016) by Kieron Connolly, and *Abandoned Places: 60 Stories of Places Where Time Stopped* (2015) by Richard Happer.
11 Tong Lam, *Abandoned Futures: A Journey to the Posthuman World* (Darlington: Carpet Bombing Culture, 2013). See also his website, www.tonglam.com/.
12 Due to the difficulties and cost of obtaining permissions for photographic images, none of these is reproduced here.
13 George Steinmetz, 'Colonial Melancholy and Fordist Nostalgia: The Ruinscapes of Namibia and Detroit', in *Ruins of Modernity*, ed. by Julia Hell and Andreas Schönle (Durham, NC: Duke University Press, 2010), pp. 294–321 (p. 295).
14 United States Census Bureau, *State & County Quickfacts*, Michigan www.census.gov/quickfacts/table/PST045215/2622000 [accessed 20 October 2016].
15 Dora Apel, *Beautiful Terrible Ruins: Detroit and the Anxiety of Decline* (New Brunswick, NJ: Rutgers University Press, 2015), p. 4.
16 Joshua Atkinson and Rosati Clayton, 'DetroitYES! and the Fabulous Ruins Virtual Tour: The Role of Diffused Intertextual Production in the Construction

of Alternative Cityscapes', *Critical Studies in Media Communication*, 29.1 (2012), 45–64 (p. 46).
17 Edensor and DeSilvey, p. 46.
18 Andrew Sargus Klein, 'Poverty is so Photogenic', *Splice Today*, 3 January 2011, www.splicetoday.com/pop-culture/poverty-is-so-photogenic [accessed 4 June 2016].
19 Eli Rosenberg, 'Motown or Ghostown? Ruin Porn in Detroit', *The Wire*, 20 January 2011, www.thewire.com/entertainment/2011/01/motown-or-ghostown-ruin-porn-in-detroit/21443/ [accessed 4 June 2016].
20 Steinmetz, pp. 295–296.
21 Strangleman, p. 25.
22 Elizabeth Blackmar, 'Modernist Ruins', *American Quarterly*, 53.2 (2001), 324–339.
23 Edensor and DeSilvey, p. 468.
24 Tim Edensor, 'Mundane Hauntings: Commuting Through the Phantasmagoric Working-Class Spaces of Manchester, England', *Cultural Geographies*, 15.3 (2008), 313–333.
25 Edensor, 'Mundane Hauntings', p. 313.
26 Edensor, 'Mundane Hauntings', p. 319.
27 Justin Armstrong, 'Everyday Afterlife: Walter Benjamin and the Politics of Abandonment in Saskatchewan, Canada', *Cultural Studies*, 25.3 (2011), 273–293 (p. 274).
28 Armstrong, p. 288.
29 Benjamin, *The Arcades Project*, p. 83.
30 Benjamin, *The Arcades Project*, p. 84.
31 Benjamin, *The Arcades Project*, p. 84.
32 Vergara, *American Ruins*, p. 11.
33 Vergara, *American Ruins*, p. 28.
34 Joseph D. Lewandowsky, 'Rescuing Critique: On the Ghetto Photography of Camilo Vergara', *Theory, Culture and Society*, 25.7–8 (2008), 285–308 (p. 286).
35 Lewandowsky, p. 286.
36 Lewandowsky, p. 287.
37 Marc Augé, *Non Places: An Introduction to Supermodernity* (London: Verso, 1995).
38 Augé, p. 78.
39 Augé, p. 75.
40 Augé p. 76.
41 Augé, p. 63.
42 Augé, p. 76.
43 Augé, p. 78.
44 Augé, p. 83.
45 Augé, p. 83.
46 Camilo José Vergara, *The New American Ghetto* (New Brunswick, NJ: Rutgers University Press), 1995, p. 69. Lewandowsky analyses the same set of images, pp. 292–295.
47 Klein, 'Poverty is so Photogenic'.
48 Matthew Newton, 'Last Exit Detroit: Christian Burkert's View of Motor City', *Thought Catalogue*, 1 December 2010, http://thoughtcatalog.com/matthew-newton/2010/12/last-exit-detroit-christian-burkerts-view-of-motor-city/ [accessed 8 June 2016].
49 J. Gordon Rodwan and John G. Rodwan, *Detroit Is: An Essay in Photographs* (Staatsburg, NY: KMW Studio, 2015), p. 15.
50 Sontag, *On Photography*, p. 23; John Berger, *Ways of Seeing* (London: Penguin, 2008), p. 29.
51 Leary, 'Detroitism'.

52 Vergara, *American Ruins*, pp. 35–39.
53 Lam, *Abandoned Futures*.
54 Most of his chapter titles indirectly refer to the type of structures in the images. For example, 'Maximizing Life' includes photographs of abandoned former hospitals and sanatoriums, and 'Faith' includes photographs of abandoned churches. Interestingly, the only chapter title with a clear geographical focus is 'Detroit' which is evidently deemed a topic on its own.
55 Leary, 'Detroitism'.
56 Leary, 'Detroitism'.
57 Sontag, *On Photography*, pp. 22–23.
58 Jameson, '*The Shining*', p. 129.
59 Leary, 'Detroitism'.
60 Richard B. Woodward, 'Disaster Photography: When is Documentary Exploitation', *Art News*, 2 June 2013, www.artnews.com/2013/02/06/the-debate-over-ruin-porn/ [accessed 15 June 2016].
61 Van Rensbergen, p. 72.
62 Julian Temple, 'Detroit: The Last Days', *Guardian*, 10 March 2010, www.theguardian.com/film/2010/mar/10/detroit-motor-city-urban-decline [accessed 13 September 2016].
63 Klein, 'Poverty is so Photogenic' (my italics).
64 Apel, p. 20.
65 Jameson, '*The Shining*', p. 116.
66 John Cunningham, 'Boredom in the Charnel House: Theses on "Post-industrial Ruin"', *Variant* 42 (2011), www.variant.org.uk/42texts/CharnelHouse.html [accessed 3 September 2016].
67 Cunningham, 'Boredom in the Charnel House'.
68 Benjamin, 'The Image of Proust', *Illuminations*, p. 211.
69 Benjamin, 'Theses on the Philosophy of History', *Illuminations*, p. 255.
70 Edensor, 'Mundane Hauntings', pp. 325, 326 (my italics).
71 Gross, pp. 133–134.
72 Fisher, *Ghosts of My Life*, p. 14.
73 Benjamin, 'On Some Motifs in Baudelaire', *Illuminations*, p. 163.
74 Van Rensbergen, pp. 92–96.
75 Jameson, '*The Shining*', p. 122.
76 Grainge, p. 30.
77 Kyle Chayka, 'Detroit Ruin Porn and the Fetish for Decay', *Hyperallergic*, 13 January 2011, http://hyperallergic.com/16596/detroit-ruin-porn/ [accessed 20 September 2016].
78 Rosenberg, 'Motown or Ghostown? Ruin Porn in Detroit'.
79 Apel, p. 5.
80 Apel, p. 2.
81 Jameson, '*The Shining*', p. 124.
82 DeSilvey and Edensor, p. 471.
83 Bloch, *Heritage of Our Times*, p. 108.
84 *The Pervert's Guide to Ideology*, dir. by Slavoj Žižek (Zeitgeist Films, 2012).
85 United States Census Bureau, *State & County Quickfacts*, Michigan.
86 Jameson, *Postmodernism* (1992), p. 40.
87 J. M. Coetzee, 'The Picturesque, the Sublime, and the South African landscape', in *White Writing: On the Culture of Letters in South Africa* (New Haven, CT: Yale University Press, 1988), p. 52.
88 Jean Baudrillard, *America*, trans. by Chris Turner (London: Verso, 2010).

89 Baudrillard, *America*, p. 6.

90 In distinguishing between Baudrillard's use of the desert to encapsulate an aesthetic associated with a certain kind of America (which I have equated, broadly, with Augé's 'non-place'), and any discussion of the aesthetic of the actual American desert, it is important to note that the latter has its own history and is often associated with a kind of American sublime. Coetzee, for example, in the same essay in which he comments on the verticality of the European sublime, observes that:

> The spaciousness, grandeur and sublimity of American landscape art, and the linking of physical expanses with expansiveness of souls, feeds and is fed by the popular conviction that American space is the natural environment of a race fitted for a spacious destiny.
>
> (p. 62)

91 Baudrillard, *America*, p. 11.

92 Baudrillard, *America*, p. 30.

93 Baudrillard, *America*, p. 22.

94 Wenders' work often exhibits the aesthetics of abandonment, so his work is not out of place in this chapter. Wim Wenders' *Places, Strange and Quiet* (Ostfildern: Hatje Cantz, 2013) is a book of photography with many images of abandoned places, a theme he took up again in his 2015 exhibition at the Blain Southern gallery in Berlin, titled *Time Capsules: By the Side of the Road.*

95 This title *Paris, Texas*, also can be read in this vein. Paris, the European city of which the Eiffel tower is the iconic image is re-made in Texas, a state typified by vast open deserts. The vertical (Paris, France) becomes, in some sense, horizontal (Paris, Texas).

96 Baudrillard, *America*, p. 9.

Conclusion

The expression 'pan out', meaning to 'end up, conclude', is derived from panning for gold during the California Gold Rush (Oxford English Dictionary). Given that my own, Bloch-inspired, approach to contemporary nostalgia culture has involved digging for gold in the ideological rubble, it is worth now standing back and asking: How much gold have I found, what is its quality, and what is it now worth? Are there any broader insights that can be drawn from these three in-depth readings of contemporary American nostalgia texts?

In Tavi Gevinson's dreamy, kitsch nostalgia for an imagined America that she has come to know primarily via filmic representation, and in her efforts to salvage the real America on her doorstep via a practice of collecting guided by the principle of 'strange magic', I discovered a profound longing. Far from merely reflecting a depthless historicity or superficial attraction to outdated styles, I argued that Tavi's nostalgia is on some level a nostalgia for the Real: for the forgotten potentialities of authenticity, sincerity, belief and wonder that she, as a child of Baudrillard's world of simulacra, no longer finds easily accessible. Tavi's metamodern affect reflects the struggle of these lost possibilities to find a new language in a postmodern culture that has all but erased them. I also read a strong resistance to instrumental reason in her archivist memory practice, a desire to redeem a suburban existence that might otherwise represent a triumph of the deadened logic of instrumentality. These resistances and hopes are frustrated, to be sure. They do not present themselves as fully formed utopian ideals. But they are the stirrings of ghosts that still haunt the cultural imaginary, demanding attention.

These 'ghosts which need redeeming' are the same ghosts, albeit in slightly different shapes, that haunt the narrator(s) in, and viewers of, *The Virgin Suicides*, explaining in part the nostalgic appeal of the film to Tavi and her fellow Rookies. I explored, in particular, the ghost of *Erfahrung* that the teenage ideal signifies, as opposed to the mere *Erlebnis* that typifies adulthood, and how, in this, too, there lies a powerful resistance to instrumental reason. The narrators in the film have *still not* forgotten Bloch's 'inconstructable question', and the film asks its viewers not to either. Moreover, the teenage nostalgia of *The Virgin Suicides*, and of other 1990s nostalgia films such as *Dazed and Confused*, functions as a kind of nostalgia for the possibility of rebellion, or negation. If nostalgia for childhood arose in part as a resistance to industrialisation and contained the forgotten utopian promise of innocence and nature, the widespread nostalgia for the Teenage in post-Cold War America could be seen as a longing for the possibility of rebellion against 'what is'. This possibility is a necessary premise of utopian thinking that has receded from grasp in a society in which the diffuseness of power seems only to confirm the sense that 'there is no alternative'.

In considering the popular genre of Detroit 'abandoned spaces' photography, I questioned readings that understand the images primarily as expressing a restorative nostalgia for Fordian prosperity. Acknowledging that these images are consumed in aestheticised passivity, I nevertheless read in both their production and consumption a powerful latent nostalgia for the vertical: for the deep sense of time, history and the future that was still accessible in the Baudelairean city but recedes from view in the perpetual present of non-place, the 'deserts of meaninglessness' in which no before and no after are visible. In this nostalgia for the vertical there is a powerful resistance to the lack of a sense of real history in late capitalism, and a desire for the possibility of a tomorrow that is distinct from today. I also read, in the allure of abandonment that these images express, the same resistance to utilitarian logic and instrumental reason that was evident in the previous two texts: the photographs stage an encounter with uselessness, with that which the logic of consumerism damns or impoverishes, which is itself a form of resistance to that logic.

These 'strange twinklings' and 'pregnant echoes' are intimations of a different way of being in the world and they recall Bloch's warning and call to arms:

> All the more does that fog descend, and that acute loss of certainty [...]. All around us also rules the devil of the cold again [...] who can best rule undisturbed as pure nothingness, as complete demystification, barring the mystery from us. But out of this simultaneously arises the paradoxical courage to prophesy the light precisely out of the fog, or in other words: *the No could not be so strong if there were not among us, at the same time, a dangerous and battle-worthy Yes.*[1]

A website that guides amateur gold panners cautions that it is a mistake, when panning for gold, to be

> too rough with your pan, which causes small flakes of gold to wash away in the water. While you may still have some of the larger flakes and nuggets left in your pan, the small ones will be carried away by the river's current.[2]

I want, similarly, to caution that in summarising the key arguments of my three readings I do not wish to wash over, or let go, the many smaller insights. The moments of nuance and critical acuity that I hope the chapters have offered make sense of contemporary nostalgia culture in a way that a short summation of these analyses can not. However, a reminder of some of the broader claims of each chapter is helpful as it allows us to stand back and ask what all three readings have in common.

In each, I argued that texts which might at first be understood as exemplary of Jameson's nostalgia mode in fact express deeper longings and tussles with modernity (and, indeed, postmodernity) than they seem to at first. This mode-ish nostalgia culture is not only ideological, a depthless consumer mode, but contains within it powerful, if muted, resistances to contemporary ideology. Sometimes these desires could be described as containing latent utopian content, as for example in the resistance to instrumental reason evident in all three texts. Other times, the veiled desires that motivate these nostalgia texts suggest a longing for the mere possibility of utopian thinking. For example, both the desire for the vertical (for a tomorrow that is distinct from yesterday) and the desire for negation (for the conceptual possibility of an alternative to 'what is') are premises of any kind of utopianism (see Chapter 2), rather than specific dreams for the future.

According to my readings, the nostalgia 'mode' is less distinct from the nostalgia 'mood' than Jameson's initial diagnosis suggests. The nostalgia culture that is so widespread across visual media today is not merely a neurotic historicism in a society with no access to history, nor is it only a superficial fashion resulting from new technologies. Although both of these positions have some merit, they do not do justice to what is at stake in contemporary nostalgia culture, that is, to the real and deep longing that it fuels and feeds off, and to the hopes and critical resistances contained within that longing. Nor does viewing these kinds of contemporary nostalgia as merely a 'mode' allow for a serious consideration of the objects of nostalgia, the possibility that, although the nostalgia has little specific historical content, it might nonetheless point towards specific, and still unfulfilled, promises of the past. In other words, I have argued that the content of the nostalgia wish is worthy of more reflection than Jameson allows and that nostalgia is not always, or only, a sign of stuck-ness or of an inability to engage with the present or future; on the contrary, sometimes it signals a refusal to give up on something important. That 'something' therefore merits considered analysis, not quick dismissal as a superficial stylised pastness with no real content. For example, in *The Virgin Suicides* the tendency of critics to focus on the stuck-ness of the grown narrators in some kind of vague childhood misses the particularities of that for which they are nostalgic: the teenage years, which are associated with very particular desires (in particular for *Erfahrung* and for the possibility of negation), the appeal of which can only be understood in light of present-day frustrations, just as the emerging appeal of childhood in the nineteenth century could only be understood in the context of rapid industrialisation.

Jameson's and Bloch's approaches are not, however, entirely distinct. That Jameson understands and appreciates Bloch's contribution to literary and cultural criticism more broadly is evident in his chapter on Bloch's literary criticism in

Marxism and Form. It is true that this broad appreciation did not have much impact on his critique of the nostalgia mode, which adopts a traditional approach to cultural criticism – exposing the ways in which cultural products reflect the dominant ideology – rather than a hermeneutic approach as advocated by Bloch. But there is another way of looking at the distinction between a Jamesonian and Blochian approach to nostalgia, which I will now examine by turning one last time to Jameson's multi-faceted reading of *The Shining*.

As discussed in Chapter 5, in his layered reading of *The Shining*, Jameson first posits the more overt content of this nostalgia film as a kind of longing for collectivity, for the 'knowable community' as embodied in the extended family. He comments that 'the disappearance of the grandparents from an atomized suburban culture must then have a significant effect on the social amnesia, the loss of a sense of the past, in a consumer society'.[3] This is reminiscent of Adorno and Horkheimer's point that:

> The bourgeois family name which [...] once individualized its bearer by relating him to his own past history seems antiquated. It arouses a strange embarrassment in Americans. In order to hide the awkward distance between individuals they call one another 'Bob' and 'Harry,' as interchangeable team members.[4]

'Bob' and 'Harry' are exactly the kind of figures one might imagine inhabiting the now characterless, corporate Overlook Hotel in *The Shining*, and Jameson's point is that the film expresses a drive towards community that it can only locate in the past. However, by locating this past specifically in the 1920s, Jameson suggests that the film exposes this longing for community or collectivity (an understandable, hopeful desire) as something more sinister: at heart it is an ideological desire for rigid class structure. Jameson's reading of *The Shining* is a dialectical reading. First, he names the expressly utopian content of the film (desire for community). Then he exposes the ideological content that underlies this (desire for rigid class structure). In doing so, he opens up more genuine utopian possibilities. The film goes back to the 1920s in order to highlight the latent desire for overt class structure, but it thereby also opens up the possibility of class consciousness. The moment that class is out in the open is the only moment that class consciousness can arise and that a sense of history – a tomorrow distinct from today – is possible. 'Bob' and 'Harry' cannot remake themselves, because they have no idea who they are, just as driving through a non-place gives little sense of the history that came before, or therefore, crucially, of what might come after: it rather suggests an eternal present in which everything is as it must be.

Bloch's reading of the 'secret Germany' of the Weimar years is also a dialectical reading, only it begins from a different point. Rather than starting with the utopian content and revealing the ideological content, Bloch begins by describing the ideological content of the cultural nostalgia of that era ('an illogical space [...] in which wishes and romanticisms, primeval drives and mythicisms revive') but then he unearths latent utopian content as well (resistance to rationalisation and to the cold rule of technology; desire for more whole relations between human beings).[5]

The point is that both Bloch and Jameson read utopian and ideological content in these respective examples of nostalgia. However, Jameson sees it as more pressing to expose the ideological content of the nostalgia mode, whereas Bloch sees it as more pressing not to dismiss subjectively non-contemporaneous currents entirely as ideology, and rather to recognise the latent utopian content that they also contain. I propose that neither direction of analysis (utopian→ideological; ideological→utopian) is inherently better than the other. What requires unearthing or exposing by the critic depends on the perspective and historical moment within which she or he is working. When Jameson first wrote about the nostalgia mode, and about postmodern culture more broadly, it had not yet been properly understood, and its ideological content had not yet been exposed. By now, in critical and cultural theory, the critiques of the nostalgia mode are so well understood that they have begun to seem inadequate in characterising such a complex phenomenon, and the imperative to search for the gold within the rubble has become more urgent.

There is also the broader function of nostalgia in the American, and indeed global, cultural context to consider. My own object of study has been nostalgia in literary and popular culture. The kinds of nostalgia I have explored evoke a dreamy, filmic, often teenage, suburban (or abandoned urban) America: neoromantic, metamodern, kitsch. I began by observing a particular kind of nostalgia that was fairly widespread in contemporary culture, and set out to better understand it. I hope that I have done this. However, one cannot write a book about contemporary nostalgia and fail to notice the myriad ways in which nostalgia is a motivating cultural force at this time, many of which are darker and more potent than the nostalgia cultures that I have explored. Donald Trump's election to office as President of the United States of America, and Britain's vote to leave the European Union, are both events that have prompted many critics to observe, and bemoan, the role of a pernicious kind of cultural nostalgia.

Bloch, in *Heritage of Our Times*, sought to better understand the kinds of cultural nostalgia that were brewing in Weimar Germany. In an essay called 'Rage and Merriment' written in 1929 he describes a marathon all-night dance championship in Frankfurt. This spectacle of light entertainment was hardly an obvious red flag (there were many more overtly worrying signs) but in it Bloch saw 'nastiness and brutality […] stupidity and ignorance' and he warns: '*What the soul of the people is cooking up here will shortly be served up in quite an acceptable form.*'[6] I do not mention this in order to draw any straightforward or trite comparisons between the current political situation and that of Nazi Germany. I mention it because Bloch's question – what is the soul of the people cooking up? – is one which many people today are asking themselves.

The particular kinds of contemporary nostalgia culture I have explored are very different in kind to the cultural nostalgia that motivated and sustained

Donald Trump's campaign ('make America great again!') and that help to explain its widespread appeal among certain voters. I am reminded of Svetlana Boym's observation:

> The first decade of the twenty-first century is not characterized by the search for newness, but by the proliferation of nostalgias that are often at odds with one another. Nostalgic cyberpunks and nostalgic hippies, nostalgic nationalists and nostalgic cosmopolitans, nostalgic environmentalists and nostalgic metrophiliacs (city lovers) exchange pixel fire in the blogosphere.[7]

But despite the different values and hopes embedded in different kinds of nostalgia, I think that a Blochian method can help us better understand and engage with *all* kinds of nostalgia, even those which seem the very worst. In Bloch's analysis of the subjectively non-contemporaneous currents of Weimar Germany he recognised, as did his fellow Marxists at the time, the petty insecurities and dangerous hatreds that the romantic blood-and-soil myths reflected and stirred up. But he insisted that we sift out from within these sentiments the 'fields of a different irratio' that were *also* entangled with these nostalgic longings of the 'average man', and emphasised the need to speak to these.[8] His point was that regression and hope do not always announce themselves clearly; nostalgia and utopia are not always clearly delineated. Or, as he puts it in his debate with Lukács:

> one must ask: *are there not dialectical links between growth and decay?* Are confusion, immaturity and incomprehensibility always and in every case to be categorized as bourgeois decadence? Might they not equally – in contrast with this simplistic and surely unrevolutionary view – be part of the transition from the old world to the new?[9]

I want at this point to raise an important question: What is the role of the critic, and what is the role of the political activist? In this book I have given three detailed readings of contemporary nostalgia culture in which I have identified latent resistances to present-day ideology – to instrumental reason and to the felt maxim that 'there is no alternative' – hiding in plain sight within these instances of a seemingly superficial nostalgia mode. But what do those resistances amount to? If they coexist happily within all-encompassing late capitalism then are they anything other than Bloch's 'abstract utopia', or 'booty for swindlers'?[10] Do they lack the transformative potential that is required for something to count as properly utopian?

As I pointed out in Chapter 3, the aspects that Bloch identifies as hopeful within the otherwise highly problematic cultural nostalgia in Weimar Germany could also be accused of lacking the transformative potential that he insists is a necessary condition of concrete utopia. But in practice, Bloch is more favourably inclined towards what may seem to be mere 'wishing' (as opposed to more active 'wanting') and is highly critical of those who dismiss wishful thinking, arguing that 'it sounds like a eunuch accusing the infant Hercules of impotence'.[11] If something is sufficiently resistant to 'what is' then, even if it is muffled, it could be said to contain latent transformative potential.

And this is where, I think, the work of political activism comes in. It is the work of the critic to identify hope-content in the cultures of our moment. It is the work of the activist to successfully appeal to those utopian shards, to help transform mood into matter. But the work of the critic, particularly the critic of popular culture, is still pressing. Bloch's analyses, for example, help us to better understand the longings and hopes driving even the most seemingly banal popular cultures of his moment. If popular culture is a key to the mood of a historical moment, then this understanding is hugely important. If the mood of the moment is not understood, it cannot be appealed to, as evidenced in the 2016 American election. Moreover, if the latent hopes are not heard and addressed, then the more readily available fears will be.

The strange and sometimes problematic fact about doing contemporary cultural criticism is that it always threatens to become outdated, to lose touch with the present. Indeed, that which you are working on can change even while you are working on it. As I mentioned in Chapter 4, Tavi Gevinson's media output has shifted significantly in philosophy and tone since she left school and moved to New York. Imagine my consternation, as I write about the nostalgia that pervades Tavi Gevinson's work, when I read the following in her 2016 editor's letter for *Rookiemag*:

> I have gotten so much moony purply melancholic mileage out of the UV drip of sadness that comes from replaying and reflecting instead of going out and discovering. But I was once told that 'nostalgia is a yearning for a time we know we can survive.' It can *feel* transporting, but it ultimately aims to seek only the familiar. It seems edgy and dark to wallow and resign, but it's also cowardly: I know exactly what I'm going to feel when I do it. What's the new experience there? Where's the bravery? I know hope and love are corny, but they are actually so much more lionhearted than the alternatives because they require engaging with the unknown, and choosing to see its majesty.[12]

At the very moment that I am putting the finishing touches to my argument that Tavi Gevinson's dreamy nostalgia is in fact more hopeful than it first seems, she decides to repudiate her nostalgic output, damning it in precisely those tired terms – of nostalgia against hope – that I have tried so hard to complicate. On the one hand, I do not think this matters: my goal was not to understand the personal motivation or psychology of this one person, but to ask why her mode of approaching the past and present appealed so widely and what hope-content her cultural output contained. But on the other hand, it does highlight some of the complications that face the critic of contemporary culture.

A more far-reaching and complex aspect of our contemporary moment that is in flux is the status of the maxim that 'there is no alternative'. Many theorists and critics of the post-Cold War era have agreed that, while the hold of this creed is enormously problematic and demands to be challenged, its truth is widely presumed. It is as a resistance to the oppressive weight of this maxim and its grip on contemporary culture that the allure of the Teenage was explained as an underlying desire for negation, or even for the conceptual possibility of 'a space outside' what currently is. However, even during the years that I worked first on my PhD and subsequently on this book, shifts have occurred that threaten to stir up the dominant perceptions that have governed the make-up of the post-Cold War western world since 1989. Radical, militant Islamism is not an appealing alternative to 'what is', but it is an alternative. Likewise, the populist nationalisms on the horizon are of a different quality to the dominant value systems of the last quarter century. These alternatives are far from utopian, at least from this author's perspective, but they do put a dent in the felt sense that the existing world order is an irrefutable fact.

As we have been taught by both Raymond Williams, in his analysis of residual and emergent cultures, and Bloch, in his analysis of non-contemporaneity, culture is not all of a piece: it does not move forward homogeneously at the front of history. Any analysis of particular cultural practices, like my own, must therefore be quite focused in its claims. If the analyses speak to truths of the broader cultural moment, as I believe mine do, then it is better not to stretch them too far. One should not make absolute or definitive claims about the culture as a whole, and risk reifying as abstractions observations about matters which are by necessity fluid. Popular culture is too mutable, and too staggered, to be tied down in abstraction. This, I think, is one of the advantages of the method I have used: analysing points in a contemporary cultural constellation that has a real, but not static, existence.

It may also mean that the lasting value of my work here is likely to lie in its demonstration of an approach to cultural criticism, to popular culture, and to reading nostalgia, which will continue to have merit, even as the particular constellation of nostalgia culture that I have explored inevitably evolves and shifts. What I have claimed of contemporary American cultural practices is that the nostalgia mode is not as divorced from the nostalgia mood as Jameson's famous analysis suggests, that despite its superficial stylised surface, it too is a reflection of deep longings and tussles with time, albeit tussles that are both veiled and frustrated. I have claimed, rather, that nostalgia is often the container of conflicted desires. In the already discussed and well-known scene in *Mad Men* when Don Draper presents his ideas to Kodak, he says:

> Nostalgia – it's delicate, but potent. Teddy told me that in Greek, 'nostalgia' literally means the pain from an old wound. It's a twinge in your heart far more painful than memory alone. This device isn't a spaceship, it's a time machine. It goes backwards, and forwards ... it takes us to a place

> where we ache to go again. It's not called the wheel, it's called the carousel. It lets us travel the way a child travels – around and around, and back home again, to a place where we know we are loved.[13]

Backwards, and forwards: nostalgia can contain *both* regressive desires for an imagined certainty *and* something else, ghosts which have not been laid to rest but continue to haunt us, humming a quietly unsettling dis-contemporaneous tune that jars against our complicity with the present. This latter part, this gold within the rubble, is not nostalgia for what once was, but nostalgia for what is *still not* ... *Not Yet*, for the unfinished promises of yesterday which continue to insist, demanding our attention, even if only in our dream-intoxicated nights. These latent desires and resistances to 'what is', these barely conscious but none-the-less powerful informants of longing, these are the 'life-light, containing nothing stale' that 'shines vexatiously', 'the freshness, the otherness', the dream 'that does not stop inserting itself into the gaps': hope.[14]

Notes

1 Hammond, pp. 15–16; Bloch, *The Spirit of Utopia*, p. 171.
2 '3 of the Biggest Mistakes Made when Panning for Gold', Luckypanner.com, www.luckypanner.com/3-of-the-biggest-mistakes-made-when-panning-for-gold/ [accessed 11 November 2016].
3 Jameson, '*The Shining*', pp. 120–121.
4 Adorno and Horkheimer, p. 165.
5 Bloch, *Heritage of Our Times*, p. 102.
6 Bloch, *Heritage of Our Times*, p. 43 (my italics).
7 Svetlana Boym, 'Nostalgia and its Discontents', *The Hedgehog Review*, 9.2 (2007), 7–18 (p. 17).
8 Bloch, *Heritage of Our Times*, p. 103.
9 Ernst Bloch, 'Discussing Expressionism', in *Aesthetics and Politics*, ed. by Ernst Bloch *et al.* (London: Verso, 2007), pp. 16–28 (p. 23).
10 Bloch, *The Principle of Hope*, I, p. 3.
11 Bloch, *The Principle of Hope*, I, p. 145.
12 Tavi Gevinson, 'Editor's Letter – November 2016: Awakening', *Style Rookie*, 1 November 2016, www.rookiemag.com/2016/11/editors-letter-61/ [accessed 1 November 2016].
13 'The Wheel', *Mad Men*, dir. by Matthew Weiner (Debmar-Mercury, 2007).
14 Bloch, *The Principle of Hope*, I, p. 28, p. 22, p. 29.

References

Abramson, Seth, 'Metamodernism: The Basics', *The Huffington Post*, 12 December 2014, www.huffingtonpost.com/seth-abramson/metamodernism-the-basics_b_5973184.html [accessed 14 July 2016].

Adorno, Theodor W. and Benjamin, Walter, *The Complete Correspondence 1928–1940* (Cambridge, MA: Harvard University Press, 1999).

Adorno, Theodor W. and Horkheimer, Max, *Dialectic of Enlightenment*, trans. by John Cumming (London: Verso, 2010).

Adorno, Theodor W., 'Cultural Criticism and Society', in *Prisms*, by Theodor W. Adorno, trans. by Samuel and Shierry Weber (Cambridge, MA: MIT Press, 1983), pp. 17–35.

Adorno, Theodor W., *Notes to Literature*, vol. 1, trans. by Shierry Weber Nicholsen (New York: Columbia University Press, 1991).

Agamben, Giorgio, *What is an Apparatus?, and Other Essays*, trans. by David Kishik and Stefan Pedatella (Stanford: Stanford University Press, 2009).

Apel, Dora, *Beautiful Terrible Ruins: Detroit and the Anxiety of Decline* (New Brunswick, NJ: Rutgers University Press, 2015).

Appadurai, Arjun, *Modernity at Large: Cultural Dimensions of Globalization* (Minneapolis: University of Minnesota Press, 1996).

Arendt, Hannah, 'Introduction: Walter Benjamin: 1892–1940', in *Illuminations: Essays and Reflections*, by Walter Benjamin, trans. by Harry Zohn (New York: Schocken Books, 1969), pp. 1–59.

Armstrong, Justin, 'Everyday Afterlife: Walter Benjamin and the Politics of Abandonment in Saskatchewan, Canada', *Cultural Studies*, 25.3 (2011), 273–293.

Atkinson, Joshua and Clayton, Rosati, 'DetroitYES! and the Fabulous Ruins Virtual Tour: The Role of Diffused Intertextual Production in the Construction of Alternative Cityscapes', *Critical Studies in Media Communication*, 29.1 (2012), 45–64.

Augé, Marc, *Non Places: An Introduction to Supermodernity* (London: Verso, 1995).

Bartholeyns, Gil, 'The Instant Past: Nostalgia and Digital Retro Photography', in *Media and Nostalgia: Yearning for the Past, Present and Future*, ed. by Katharina Niemeyer (London: Palgrave Macmillan, 2014), pp. 51–69.

Baudrillard, Jean, *America*, trans. by Chris Turner (London: Verso, 2010).

Baudrillard, Jean, *Simulacra and Simulation*, trans. by Sheila Faria Glaser (Ann Arbor: University of Michigan Press, 1994).

Bauman, Zygmunt, *Liquid Modernity* (Cambridge: Polity Press, 2000).

Bauman, Zygmunt, *Liquid Times: Living in an Age of Uncertainty* (Cambridge: Polity Press, 2007).

Baxter, Hannah, 'Tavi Gevinson Reminds Us Why She's The World's Coolest Teenager', *Bust*, http://bust.com/general/12826-tavi-gevinson-reminds-us-why-she-is-the-worlds-coolest-teenager.html [accessed 4 August 2014].

Beckett, Charles, 'How To Think About The Future', www.howtothinkaboutthefuture.com/?p=75 [accessed 10 January 2016].

Bedford, Martyn, 'Nostalgia Ain't What It Used To Be', *Literary Review*, 299 (2003).

Benjamin, Walter, *The Arcades Project*, trans. by Howard Eiland and Kevin McLaughlin (Cambridge, MA: Harvard University Press, 2002).

Benjamin, Walter, 'Experience', in *Selected Writings*, ed. by Marcus Bullock and Michael W. Jennings, vol. 1 (Cambridge, MA: Harvard University Press, 1996), pp. 3–6.

Benjamin, Walter, *Illuminations: Essays and Reflections*, trans. by Harry Zohn (New York: Schocken Books, 1969).

Benjamin, Walter, 'The Image of Proust', in *Illuminations: Essays and Reflections*, trans. by Harry Zohn (New York: Schocken Books, 1969), pp. 201–217.

Benjamin, Walter, 'The Lives of Students', in *Selected Writings*, ed. by Marcus Bullock and Michael W. Jennings, vol. 1 (Cambridge, MA: Harvard University Press, 1996), pp. 37–48.

Benjamin, Walter, *Selected Writings*, ed. by Marcus Bullock and Michael W. Jennings, vol. 1 (Cambridge, MA: Harvard University Press, 1996).

Benjamin, Walter, 'On Some Motifs in Baudelaire', in *Illuminations: Essays and Reflections*, trans. by Harry Zohn (New York: Schocken Books, 1969), pp. 155–201.

Benjamin, Walter, 'Theses on the Philosophy of History', in *Illuminations: Essays and Reflections*, trans. by Harry Zohn (New York: Schocken Books, 1969), pp. 253–265.

Benjamin, Walter, 'Unpacking My Library', in *Illuminations: Essays and Reflections*, trans. by Harry Zohn (New York: Schocken Books, 1969), pp. 59–69.

Beuka, Robert, *SuburbiaNation: Reading Suburban Landscape in Twentieth-Century American Fiction and Film* (New York: Palgrave Macmillan, 2004).

Blackmar, Elizabeth, 'Modernist Ruins', *American Quarterly*, 53.2 (2001), 324–339.

Bloch, Ernst *et al.*, *Aesthetics and Politics* (London: Verso, 2007).

Bloch, Ernst, *Atheism in Christianity*, trans. by J. T. Swann (London: Verso, 2009).

Bloch, Ernst, 'Discussing Expressionism', in *Aesthetics and Politics*, ed. by Ernst Bloch *et al.* (London: Verso, 2007), pp. 16–28.

Bloch, Ernst, *Heritage of Our Times*, trans. by Neville and Stephen Plaice (Cambridge: Polity Press, 1991).

Bloch, Ernst, *A Philosophy of the Future*, trans. by John Cumming (New York: Herder and Herder, 1970).

Bloch, Ernst, *The Principle of Hope*, trans. by Neville Plaice, Stephen Plaice and Paul Knight, 3 vols (Cambridge, MA: MIT Press, 1986).

Bloch, Ernst, *The Spirit of Utopia*, trans. by Anthony A. Nassar (Stanford: Stanford University Press, 2000).

Bloch, Ernst, *Traces*, trans. by Anthony A. Nassar (Stanford: Stanford University Press, 2006).

Bloch, Ernst, *The Utopian Function of Art and Literature: Selected Essays*, trans. by Jack Zipes and Frank Mecklenburg (Cambridge, MA: MIT Press, 1988).

Bourriard, Nicholas, *Altermodern: Tate Triennial 2009* (London: Tate Publishing, 2009).

Boym, Svetlana, *The Future of Nostalgia* (New York: Basic Books, 2001).

Boym, Svetlana, 'Nostalgia and its Discontents', *The Hedgehog Review*, 9.2 (2007), 7–18.

Boym, Svetlana, 'Nostalgic Technology: Notes for an Off-Modern Manifesto', www.svetlanaboym.com/manifesto.htm [accessed 2 November 2016].

Bridle, James, 'Hauntological Futures', http://booktwo.org/notebook/hauntological-futures/ [accessed 6 November 2019].

Chayka, Kyle, 'Detroit Ruin Porn and the Fetish for Decay', *Hyperallergic*, 13 January 2011, http://hyperallergic.com/16596/detroit-ruin-porn/ [accessed 6 November 2019].

Chopra-Gant, Mike, *'The Waltons': Nostalgia and Myth in Seventies America* (London: I. B. Tauris, 2013).

Christopher, Matthew, 'Abandoned America: An Autopsy of the American Dream', www.abandonedamerica.us/about.html [accessed 10 September 2016].

Coetzee, J. M., 'The Picturesque, the Sublime, and the South African landscape', in *White Writing: On the Culture of Letters in South Africa* (New Haven, CT: Yale University Press, 1988).

Coon, David. R, *Look Closer: Suburban Narratives and American Values in Film and Television* (New Brunswick, NJ: Rutgers University Press, 2014).

Coontz, Stefanie, *The Way We Never Were: American Families and the Nostalgia Trap* (New York: Basic Books, 2000).

Cox, Peter, 'Why I Love ... #56: The Virgin Suicides', *The Lost Highway Hotel*, 11 October 2013, https://thelosthighwayhotel.com/2013/10/11/why-i-love-56-the-virgin-suicides/ [accessed 6 November 2019].

Cunningham, John, 'Boredom in the Charnel House: Theses on "Post-industrial Ruin"', *Variant* 42 (2011), www.variant.org.uk/42texts/CharnelHouse.html [accessed 6 November 2019].

Davis, Fred, 'Nostalgia, Identity and the Current Nostalgia Wave', *The Journal of Popular Culture*, 11.2 (1977), 414–424.

Derrida, Jacques, *Spectres of Marx: The State of the Debt, the Work of Mourning and the New International*, trans. by Peggy Kamuf (New York: Routledge, 2006).

Dickinson, J. K., 'Ernst Bloch's *The Principle of Hope*: A Review of and Comment on the English Translation', *Babel*, 36.1 (1990), 7–31.

Dika, Vera, *Recycled Culture in Contemporary Art and Film: The Uses of Nostalgia* (Cambridge: Cambridge University Press, 2003).

Dines, Martin, 'Suburban Gothic and the Ethnic Uncanny in Jeffrey Eugenides' *The Virgin Suicides*', *Journal of American Studies*, 46.4 (2012), 959–975.

Dullea, Georgia, 'Sick of the 60s, 3 Men of the 80s Try to Give Nostalgia a Bad Name', *New York Times*, 30 April 1989, www.nytimes.com/1989/04/30/style/life-style-sick-of-the-60-s-3-men-of-the-80-s-try-to-give-nostalgia-a-bad-name.html [accessed 6 November 2019].

Eagleton, Terry, *Reason, Faith and Revolution: Reflections on the God Debate* (New Haven, CT: Yale University Press, 2009).

Eagleton, Terry, *Why Marx Was Right* (Yale: Yale University Press, 2011), 68–69.

Edensor, Tim, 'Mundane Hauntings: Commuting Through the Phantasmagoric Working-Class Spaces of Manchester, England', *Cultural Geographies*, 15.3 (2008), 313–333.

Edensor, Tim and Desilvey, Caitlin, 'Reckoning with Ruins', *Progress in Human Geography*, 37.4 (2013), 465–485.

Edwards, Caroline, 'Fictions of the Not Yet: Time and the Contemporary British Novel' (unpublished doctoral thesis, University of Nottingham, 2010).

Edwards, Caroline, 'Uncovering the "Gold-Bearing Rubble": Ernst Bloch's Literary Criticism', in *Utopianism, Modernism, and Literature in the Twentieth Century*, ed. by Alice Reeve-Tucker and Nathan Waddell (London: Palgrave Macmillan, 2013), pp. 182–203.

Eliot, Thomas Stearns, 'The Waste Land', in *The Waste Land and Other Poems* (London: Faber and Faber, 1972), pp. 21–46.

Erber, Pedro, 'Contemporaneity and Its Discontents', *Diacritics*, 41.1 (2013), 28–48.

Eugenides, Jeffrey, 'The Art of Fiction no. 215, Interview by James Gibbon', *The Paris Review*, 199 (2011) www.theparisreview.org/interviews/6117/the-art-of-fiction-no-215-jeffrey-eugenides [accessed 6 November 2019].

Eugenides, Jeffrey, *The Virgin Suicides* (London: Fourth Estate, 2013).

Fienberg, Daniel, '*Fargo* Guest Bruce Campbell on the "Terror" of Playing Ronald Reagan, Shared B-Movie Background', *The Hollywood Reporter*, 9 November 2015, www.hollywoodreporter.com/live-feed/fargo-bruce-campbell-ronald-reagan-838622 [accessed 6 November 2019].

Fisher, Mark, *Capitalist Realism: Is There No Alternative?* (Ropley: O-Books, 2009).

Fisher, Mark, 'Corridors in which Time is Abolished: Notes on Last Year in Marienbad', *K-Punk*, 4 October 2005, http://k-punk.abstractdynamics.org/archives/2005_10.html [accessed 7 May 2015].

Fisher, Mark, *Ghosts of My Life: Writings on Depression, Hauntology and Lost Futures* (Alresford: O-Books, 2014).

Flinn, Caryl, *Strains of Utopia: Gender, Nostalgia, and Hollywood Film Music* (Princeton, NJ: Princeton University Press, 1992).

Gallix, Andrew, 'Hauntology: A Not-So-New Critical Manifestation', *Guardian*, 17 June 2011, www.theguardian.com/books/booksblog/2011/jun/17/hauntology-critical [accessed 6 November 2019].

Geohegan, Vincent, *Ernst Bloch* (London: Routledge, 1996).

Gevinson, Tavi, 'Editor's Letter – December 2013: Forever', *Rookiemag*, 2 December 2013, www.rookiemag.com/2013/12/editors-letter-26/ [accessed 6 November 2019].

Gevinson, Tavi, 'Editor's Letter – November 2016: Awakening', *Style Rookie*, 1 November 2016, www.rookiemag.com/2016/11/editors-letter-61/ [accessed 6 November 2019].

Gevinson, Tavi, 'Five Years', *Style Rookie*, 3 April 2013, www.thestylerookie.com/search?updated-max=2013-12-03T00:51:00-06:00&max-results=1&start=2&by-date=false [accessed 6 November 2019].

Gevinson, Tavi, 'Girls with Power and Mystique: An Interview with Sofia Coppola', www.rookiemag.com/2013/06/sofia-coppola-interview/ [accessed 6 November 2019].

Gevinson, Tavi, 'Heart n Soul', *Style Rookie*, 17 November 2011, www.thestylerookie.com/search?updated-max=2011-12-12T16:29:00-06:00&max-results=1&start=22&by-date=false [accessed 6 November 2019].

Gevinson, Tavi, 'A Hodgepodge, If You Will', *Style Rookie*, 5 January 2012, www.thestylerookie.com/2012/01/hodgepodge-if-you-will.html [accessed 6 November 2019].

Gevinson, Tavi, 'Idea's At The House: Tavi's Big Big World', YouTube, 26 August 2013, www.youtube.com/watch?v=YSkz7c4wT9A [accessed 6 November 2019].

Gevinson, Tavi, 'Moving Images: Shows and Films Imbued with Yearning', *Rookiemag.com*, June 2013, www.rookiemag.com/2013/06/movietv-rex/ [accessed 6 November 2019].

Gevinson, Tavi, 'My United States of Whatever', *Style Rookie*, 22 June 2012, www.thestylerookie.com/search?updated-max=2012-09-16T11:38:00-05:00&max-results=1&reverse-paginate=true&start=3&by-date=false [accessed 6 November 2019].

Gevinson, Tavi, 'Sixteen and Spring', *Style Rookie*, 3 May 2012, www.thestylerookie.com/search?updated-max=2012-05-20T09:28:00-05:00&max-results=1&start=5&by-date=false [accessed 6 November 2019].

Gevinson, Tavi, 'Strange Magic', *Style Rookie*, 29 December 2011, www.thestylerookie.com/search?updated-max=2012-04-01T16:08:00-05:00&max-results=1&start=8&by-date=false [accessed 6 November 2019].

Gevinson, Tavi, 'Technicolour', *Style Rookie*, 1 February 2012, www.thestylerookie.com/search?updated-max=2012-02-03T19:22:00-06:00&max-results=1&start=18&by-date=false [accessed 6 November 2019].

Gevinson, Tavi, 'A Teen Just Trying to Figure It Out', YouTube, 9 April 2012, www.youtube.com/watch?v=6osiBvQ-RRg [accessed 6 November 2019].

Gevinson, Tavi, 'Teenage Disconnect and *The Virgin Suicides*', npr, 26 December 2012, www.npr.org/2012/12/26/163570963/teenage-disconnect-and-the-virgin-suicides [accessed 6 November 2019].

Gevinson, Tavi, 'There Are Holes in the Universe', *Style Rookie*, 27 March 2012, www.thestylerookie.com/search?updated-max=2012-04-01T16:08:00-05:00&max-results=1&start=8&by-date=false [accessed 6 November 2019].

Gevinson, Tavi, 'Untitled Blog Post, *Style Rookie*, 27 March 2011, www.thestylerookie.com/2011/03/i-feel-like-photo-to-accompany-this.html [accessed 6 November 2019].

Gevinson, Tavi, 'Why The 21st Century Isn't All That Bad', *Rookiemag*, 28 October 2011, www.rookiemag.com/2011/10/21st-century-not-all-that-bad/ [accessed 6 November 2019].

Gevinson, Tavi and Collins, Petra, 'Strange Magic', *Rookiemag*, 8 October 2012, www.rookiemag.com/2012/08/strange-magic/ [accessed 6 November 2019].

Gilloch, Graeme, *Walter Benjamin: Critical Constellations* (Cambridge: Polity Press, 2002).

Gleiberman, Owen, 'Why Nostalgia Movies Leave us Dazed (but Not Confused)', *Entertainment Weekly*, 4 April 2009, www.ew.com/article/2009/04/04/why-nostalgia-m [accessed 6 November 2019].

Gonzales, Ed, '*The Virgin Suicides*', *Slant Magazine*, 2 May 2001, www.slantmagazine.com/film/review/the-virgin-suicides [accessed 6 November 2019].

Goode, Erica, 'What's In an Inkblot? Some Say, Not Much', *New York Times*, 20 February 2001, http://mobile.nytimes.com/2001/02/20/science/what-s-in-an-inkblot-some-say-not-much.html?_r=0 [accessed 6 November 2019].

Grainge, Paul, *Monochrome Memories: Nostalgia and Style in Retro America* (Westport, CT: Praeger Publishers, 2002).

Gross, David, *Lost Time: On Remembering and Forgetting in Late Modern Culture* (Boston: University of Massachusetts Press, 2000).

Gunn, Richard, 'Ernst Bloch's *The Principle of Hope*', *Edinburgh Review*, 76 (1987), 1–9.

Guttridge, Peter, 'Book Review: Dying to Escape Big Bad Mum: *The Virgin Suicides* – Jeffrey Eugenides', *The Independent*, 18 June 1993, www.independent.co.uk/arts-entertainment/books/book-review-dying-to-escape-big-bad-mum-the-virgin-suicides-jeffrey-eugenides-bloomsbury-1599-pounds-1492519.html [accessed 6 November 2019].

Halley, Peter, 'Notes on Nostalgia', *New Observations*, 28 (1985), www.peterhalley.com/ARTISTS/PETER.HALLEY/NOTES.ON.NOSTALGIA.FR2.htm [accessed 12 January 2016].

Hammond, Craig, 'Towards a Neo-Blochian Theory of Complexity, Hope and Cinematic Utopia' (unpublished doctoral thesis, University of Lancaster, 2012).

Harvey, David, *Spaces of Hope* (Edinburgh: Edinburgh University Press, 2000).

Hatherley, Owen, *Militant Modernism* (Winchester: O-Books, 2008).

Hatherley, Owen, *The Ministry of Nostalgia* (London: Verso, 2016).

Healey, Claire Marie, 'Deconstructing the Fashion of *The Virgin Suicides*', *Dazed*, 2015, www.dazeddigital.com/fashion/article/24730/1/deconstructing-the-fashion-of-the-virgin-suicides [accessed 6 November 2019].

Hebdige, Dick, *Hiding in the Light: On Images and Things* (London: Comedia, 1988).

Hell, Julia and Schönle, Andreas, *Ruins of Modernity* (Durham, NC: Duke University Press, 2010).

Heusser, Martin, 'Et in Arcadia Ego: The Pastoral Aesthetics of Suburbia in Jeffrey Eugenides' *The Virgin Suicides*', *Swiss Papers in English Language and Literature*, 20 (2007), pp. 175–189.

Higson, Andrew, 'Nostalgia is Not What It Used To Be: Heritage Films, Nostalgia Websites and Contemporary Consumers', *Consumption Markets and Culture*, 17.2 (2014), 120–142.

Hofer, Johannes, 'Dissertatio Medica de Nostalgia' (Basel, 1688), trans. by Carolyn Kiser Anspach, in *Bulletin of the History of Medicine*, 2 (1934), 376–391.

Hoskin, Bree, 'Playground Love: Landscape and Longing in Sofia Coppola's *The Virgin Suicides*', *Literature/Film Quarterly*, 35.3 (2007), 214–221.

Hudson, Wayne, *The Marxist Philosophy of Ernst Bloch* (London: Macmillan, 1982).

Hunter, Melissa, '*The Virgin Suicides*', *The Soul of the Plot*, 12 May 2015, https://thesouloftheplot.wordpress.com/2015/05/12/the-virgin-suicides/ [accessed 6 November 2019].

Hutcheon, Linda, 'Irony, Nostalgia and the Postmodern', in *Methods for the Study of Literature as Cultural Memory*, ed. by Raymond Vervliet and Annemarie Estor (Boston: Brill, 2000), pp. 189–207.

Huyssen, Andreas, 'Nostalgia for Ruins', *Grey Room* (2006), 6–21.

Huyssen, Andreas, *Twilight Memories: Marking Time in a Culture of Amnesia* (New York: Routledge, 1995).

Illbruck, Helmut, *Nostalgia: Origins and Ends of an Unenlightened Disease* (Evanston, IL: Northwestern University Press, 2012).

Jacobsen, Michael Hviid, 'Liquid Modern "Utopia" – Zygmunt Bauman on the Transformation of Utopia', in *Utopia: Social Theory and The Future*, ed. by Michael Hviid Jacobsen and Keith Tester (Farnham: Ashgate, 2012), pp. 69–96.

Jameson, Fredric, *Archaeologies of the Future: The Desire Called Utopia and Other Science Fictions* (London: Verso, 2005).

Jameson, Fredric, *Marxism and Form: Twentieth Century Dialectical Theories of Literature* (Princeton, NJ: Princeton University Press, 1974).

Jameson, Fredric, 'Postmodernism, or The Cultural Logic of Late Capitalism', *New Left Review*, 146 (1984), 53–92.

Jameson, Fredric, *Postmodernism, or The Cultural Logic of Late Capitalism* (London: Verso, 1992).

Jameson, Fredric, '*The Shining*', *Social Text*, 4 (1981), 114–125.

Jameson, Fredric and Stefanson, Anders, 'Regarding Postmodernism – A Conversation with Fredric Jameson', *Social Text*, 21 (1989), 3–30.

Jay, Martin, 'Anamnestic Totalization: Reflections on Marcuse's Theory of Remembrance', *Theory and Society*, 11.1 (1982), 1–15.

Jay, Martin, *The Dialectical Imagination: A History of The Frankfurt School and the Institute of Social Research, 1923–1950* (Berkeley: University of California Press, 1996).

Jenß, Heike, 'Cross-Temporal Explorations: Notes on Fashion and Nostalgia', *Critical Studies in Fashion and Beauty*, 4.1–2 (2013), 107–124.

Jolly, Nick, '*The Virgin Suicides* and the Transcendental Quietude of Young Love', *The Essential*, http://theessential.com.au/features/nostalgia-files/the-virgin-suicides-and-the-transcendental-quietude-of-young-love [accessed 6 November 2019].

Kammen, Michael, *Mystic Chords of Memory: The Transformation of Tradition in American Culture* (New York: Alfred A. Knopf, 1991).

Kant, Immanuel, 'Answer the Question: What is Enlightenment?', in *Practical Philosophy*, trans. and ed. by Mary J. Gregor (Cambridge: Cambridge University Press, 1996), pp. 11–22.

Kaufman, David, 'Thanks for the Memory: Bloch, Benjamin and The Philosophy of History', in *Not Yet: Reconsidering Ernst Bloch*, ed. by Jamie Owen Daniel and Tom Moylan (London: Verso, 1997), pp. 33–53.

Kirby, Alan, *Digimodernism: How New Technologies Dismantle the Postmodern and Reconfigure Our Culture* (New York: Continuum, 2009).

Klein, Andrew Sargus, 'Poverty is so Photogenic', *Splice Today*, 3 January 2011, www.splicetoday.com/pop-culture/poverty-is-so-photogenic [accessed 6 November 2019].

Kracauer, Siegfried, *The Mass Ornament: Weimar Essays*, trans. by Thomas Y. Levin (Cambridge, MA: Harvard University Press, 1995).

Kraus, Chris, *I Love Dick* (Los Angeles: Semiotext(e), 2006).

Lam, Tong, *Abandoned Futures: A Journey to the Posthuman World* (Darlington: Carpet Bombing Culture, 2013).

Landmann, Michael, 'Talking with Ernst Bloch: Korčula, 1968', *Telos*, 25 (1975), 165–185.

Landsberg, Alison, *Prosthetic Memory: The Transformation of American Remembrance in the Age of Mass Culture* (New York: Columbia University Press, 2004).

Larocca, Amy, 'At 18, Tavi Gevinson is a Fashion Veteran – and a Broadway Rookie', *The Cut*, 10 August 2014, http://nymag.com/thecut/2014/08/tavi-gevinson-from-fashion-to-broadway.html [accessed 6 November 2019].

Latour, Bruno, *We Have Never Been Modern*, trans. by Catherine Porter (Cambridge, MA: Harvard University Press, 1993).

Leary, John Patrick, 'Detroitism', *Guernica*, 15 January 2011, www.guernicamag.com/features/leary_1_15_11/ [accessed 6 November 2019].

Le Guin, Ursula, Acceptance Speech for the National Book Foundation's Medal for Distinguished Contribution to American Letters at the 65th National Book Awards, online video recording, YouTube, 19 November 2014, www.youtube.com/watch?v=Et9Nf-rsALk&app=desktop [accessed 6 November 2019].

Levitas, Ruth, *The Concept of Utopia* (Witney: Peter Lang, 2011).

Lewandowsky, Joseph D., 'Rescuing Critique: On the Ghetto Photography of Camilo Vergara', *Theory, Culture and Society*, 25.7–8 (2008), 285–308.

Lipovetsky, Giles, *Hypermodern Times*, trans. by Andrew Brown (Cambridge: Polity Press, 2005).

Lowenthal, David, 'Nostalgia Tells It Like It Wasn't', in *The Imagined Past: History and Nostalgia*, ed. by Christopher Shaw and Malcolm Chase (Manchester: Manchester University Press, 1989), pp. 18–33.

Lukács, Georg, *The Theory of the Novel*, trans. by Anna Bostock (Cambridge, MA: MIT Press, 1968).

Mannheim, Karl, *Ideology and Utopia: An Introduction to the Sociology of Knowledge* (London: Routledge, 1960).

Marche, Stephen, '*Dazed and Confused* was the Definitive Movie About the '90s, Not the '70s', *Esquire*, 6 March 2013, www.esquire.com/entertainment/movies/a20266/dazed-and-confused-20th-anniversary-15175511/ [accessed 6 November 2019].

Marcus, Daniel, *Happy Days and Wonder Years: The Fifties and Sixties in Contemporary Cultural Politics* (New Brunswick, NJ: Rutgers University Press, 2004).

Marcuse, Herbert, *Eros and Civilization: A Philosophical Inquiry into Freud* (London: Routledge, 1998).

Marcuse, Herbert, *One-Dimensional Man* (London: Routledge, 2002).

Marx, Karl and Engels, Friedrich, *The Communist Manifesto*, ed. and trans. by L. M. Findlay (Peterborough, ON: Broadview Press, 2004).

Mayshark, Jesse Fox, *Post-pop Cinema: The Search for Meaning in New American Film* (Westport, CT: Praeger, 2007).

Mulvey, Laura, 'The Pre-Oedipal Father: The Gothicism of *Blue Velvet*', in *Modern Gothic: A Reader*, ed. by Allan Lloyd Smith and Victor Sage (Manchester: Manchester University Press, 1996), pp. 36–65.

Munoz, José Esteban, *Cruising Utopia: The Then and There of Queer Futurity* (New York: New York University Press, 2009).

Murphy, Bernice, *The Suburban Gothic in American Popular Culture* (Basingstoke: Palgrave Macmillan, 2009).

Newton, Matthew, 'Last Exit Detroit: Christian Burkert's View of Motor City', *Thought Catalogue*, 1 December 2010, http://thoughtcatalog.com/matthew-newton/2010/12/last-exit-detroit-christian-burkerts-view-of-motor-city/ [accessed 6 November 2019].

Nichols, Thea Liberty, 'The Only Thing I Do Every Day: Tavi Gevinson', *Art 21 Magazine*, 22 April 2013, http://blog.art21.org/2013/04/22/the-only-thing-i-do-everyday-tavi-gevinson/#.WBi_JdCLTnB [accessed 6 November 2019].

Ní Dhúill, Caitríona, *Sex in Imagined Spaces: Gender and Utopia from More to Bloch* (Oxford: Legenda, 2010).

Niemeyer, Katherina, 'Introduction: Media and Nostalgia', in *Media and Nostalgia: Yearning for the Past, Present and Future*, ed. by Katharina Niemeyer (London: Palgrave Macmillan, 2014), pp. 1–23.

Onstad, Katrina, 'Is Tavi Gevinson Girl Power's Last Chance?', *The Globe and Mail*, 13 September 2012, www.theglobeandmail.com/life/is-tavi-gevinson-girl-powers-last-chance/article4543738/ [accessed 6 November 2019].

The Pervert's Guide to Ideology, dir. by Slavoj Žižek (Zeitgeist Films, 2012).

Phillips, James, 'Distance, Absence and Nostalgia', in *Descriptions*, ed. by Don Ihde and Hugh J. Silverman (Albany, NY: SUNY Press, 1985).

Phipps, Keith and Robinson, Tasha, 'Forum: *The Virgin Suicides*', *The Dissolve*, 25 June 2015, https://thedissolve.com/features/movie-of-the-week/1080-forum-the-virgin-suicides/ [accessed 6 November 2019].

Pickering, Michael and Keightley, Emily, 'The Modalities of Nostalgia', *Current Sociology*, 54.6 (2006), 919–941.

Plaice, Neville and Plaice, Stephen, 'Translator's Introduction', in Ernst Bloch, *Heritage Of Our Times* (Cambridge: Polity Press, 1991), pp. ix–xiv.

Plaice, Neville, Plaice, Stephen and Knight, Paul, 'Translator's Introduction', in Ernst Bloch, *The Principle of Hope*, 3 vols (Cambridge, MA: MIT Press, 1986), pp. xix–xxxiii.

Potter, Andrew, 'Mad Men and the end of Delicate, but Potent Nostalgia', *National Post*, 15 May 2015, www.nationalpost.com/m/wp/blog.html?b=news.nationalpost.com/arts/television/mad-men-and-the-end-of-delicate-but-potent-nostalgia [accessed 6 November 2019].

Preston, Peter, 'Death and the Maidens', *Guardian*, 21 May 2000, www.theguardian.com/film/2000/may/21/peterpreston [accessed 6 November 2019].

Rann, Jamie, 'Beauty and the East: Is it Time to Kick Our Addiction to Ruin Porn?', *The Calvert Journal*, 31 July 2014, http://calvertjournal.com/features/show/6168/power-and-architecture-ruin-porn-photography-rebecca-bathory [accessed 6 November 2019].

Reynolds, Simon, *Retromania* (London: Faber and Faber, 2012).

Ricoeur, Paul, *Lectures on Ideology and Utopia*, ed. by George H. Taylor (New York: Columbia University Press, 1986).

Rodwan, J. Gordon and Rodwan, John G., *Detroit Is: An Essay in Photographs* (Staatsburg, NY: KMW Studio, 2015).

Rookiemag submissions page, http://rookiemag.com/you [accessed 2 June 2013].

Rosenberg, Eli, 'Motown or Ghostown? Ruin Porn in Detroit', *The Wire*, 20 January 2011, www.thewire.com/entertainment/2011/01/motown-or-ghostown-ruin-porn-in-detroit/21443/ [accessed 6 November 2019].

Roth, Michael, 'Dying of the Past: Medical Studies of Nostalgia in Nineteenth-Century France', *History and Memory*, 3.1 (1991), 5–29.

Ryan, Mike, '*Dazed and Confused* Would Never Work Today', *The Huffington Post*, 10 November 2013, www.huffingtonpost.com/mike-ryan/dazed-and-confused-20-years_b_4085398.html [accessed 6 November 2019].

Salinger, J. D., *Franny and Zooey* (London: Penguin, 1994).

Sapio, Guiseppina, 'Homesick for Aged Home Movies: Why Do We Shoot Contemporary Family Videos in Old-Fashioned Ways?', in *Media and Nostalgia: Yearning for the Past, Present and Future*, ed. by Katharina Niemeyer (London: Palgrave Macmillan, 2014), pp. 39–50.

Sargeant, Lyman Tower, 'The Three Faces of Utopianism Revisited', *Utopian Studies*, 5.1 (1994), 1–37.

Schmich, Mary, 'Being Tavi's Mom: What's It Like to Be The Mother of a Teen Sensation?', *Chicago Tribune*, 06 April 2014, http://articles.chicagotribune.com/2014-04-06/news/ct-tavi-gevinson-schmich-met-0406-20140406_1_tavi-gevinson-berit-engen-steve-gevinson [accessed 6 November 2019].

Schulman, Michael, 'The Oracle of Girl World', *New York Times*, 27 July 2012, www.nytimes.com/2012/07/29/fashion/tavi-gevinson-the-oracle-of-girl-world.html [accessed 6 November 2019].

Scott, A. O., 'The Death of Adulthood in American Culture', *New York Times*, 11 September 2014, www.nytimes.com/2014/09/14/magazine/the-death-of-adulthood-in-american-culture.html [accessed 6 November 2019].

Shostak, Debra, '"A Story We Could Live With": Narrative Voice, the Reader, and Jeffrey Eugenides's *The Virgin Suicides*', *Modern Fiction Studies*, 55.4 (2009), 808–832.

Siebers, Johan, 'Ernst Bloch's Dialectical Anthropology', in *The Privatization of Hope: Ernst Bloch and The Future of Utopia*, ed. by Peter Thompson and Slavoj Žižek (Durham, NC; London: Duke University Press, 2013), pp. 61–82.

Siebers, Johan, 'The Utopian Function of Film Music', in *Marx at the Movies: Revisiting History, Theory and Practice*, ed. by Ewa Mazierska and Lars Kristensen (London: Palgrave Macmillan, 2014), pp. 46–61.

Singer, Matt, 'Keynote: *American Graffiti* and George Lucas' Nostalgia for Nostalgia', *The Dissolve*, 18 March 2014, https://thedissolve.com/features/movie-of-the-week/468-keynote-american-graffiti-and-george-lucas-nostalg/ [accessed 6 November 2019].

Sloane, Leonard, 'Nostalgia for Extinct Pop Culture Creates Industry', *New York Times*, 22 March 1970, www.nytimes.com/1970/03/22/archives/nostalgia-for-extinct-pop-culture-creates-industry.html?_r=1 [accessed 3 December 2015].

Sontag, Susan, *Against Interpretation and Other Essays* (London: Penguin, 2009).

Sontag, Susan, *On Photography* (London: Penguin, 2008).

Sprengler, Christine, *Screening Nostalgia: Populuxe Props and Technicolour Aesthetics in Contemporary American Film* (New York: Berghan, 2009).

Steiner, George, *Nostalgia for the Absolute* (Concord, ON: House of Anansi Press, 1997).

Steinmetz, George, 'Colonial Melancholy and Fordist Nostalgia: The Ruinscapes of Namibia and Detroit', in *Ruins of Modernity*, ed. by Julia Hell and Andreas Schönle (Durham, NC: Duke University Press, 2010), pp. 294–321.

Stewart, Janet, 'Breaking the Power of the Past Over the Present: Psychology, Utopianism and the Frankfurt School', *Utopian Studies*, 18.1 (2007), 21–42.

Stewart, Susan, *On Longing: Narratives of the Miniature, the Gigantic, the Souvenir, the Collection* (Durham, NC: Duke University Press, 1993).

Strangleman, Tim, '"Smokestack Nostalgia", "Ruin Porn" or Working-Class Obituary: The Role and Meaning of De-industrial Representation', *International Labor and Working-Class History*, 84 (2013), 23–37.

Tait, R. Colin, 'That 70s Sequence: Remembering the Bad Old Days in Summer of Sam', *Cinephile*, 5.2 (2009) http://cinephile.ca/archives/volume-5-no-2-the-scene/that-70s-sequence-remembering-the-bad-old-days-in-summer-of-sam/ [accessed 6 November 2019].

Tannock, Stuart, 'Nostalgia Critique', *Cultural Studies*, 9.3 (1995), 453–464.

Tarnas, Richard, *Cosmos and Psyche* (New York: Viking Penguin, 2006).

Temple, Julian, 'Detroit: The Last Days', *Guardian*, 10 March 2010, www.theguardian.com/film/2010/mar/10/detroit-motor-city-urban-decline [accessed 6 November 2019].

Thompson, Peter, 'The Frankfurt School, Part 2: Negative Dialectics', *Guardian*, 1 April 2013, www.theguardian.com/commentisfree/2013/apr/01/negative-dialectics-frankfurt-school-adorno [accessed 6 November 2019].

Thompson, Peter, 'Introduction', in *Atheism and Christianity* (London: Verso, 2009).

Thompson, Peter and Žižek, Slavoj, eds., *The Privatization of Hope: Ernst Bloch and The Future of Utopia* (Durham, NC: Duke University Press, 2013).

'3 of the Biggest Mistakes Made when Panning for Gold', Luckypanner.com, www.luckypanner.com/3-of-the-biggest-mistakes-made-when-panning-for-gold/ [accessed 11 November 2016].

TIME Staff, 'The 25 Most Influential Teens of 2014', *Time Magazine*, 13 October 2014, http://time.com/3486048/most-influential-teens-2014/ [accessed 6 November 2019].

Tönnies, Ferdinand, *Community and Society (Gemeinschaft und Gesellschaft)*, trans. by Charles P. Loomis (New York: Harper & Row, 1963).

Touré, 'Nostalgia: Our Favourite Cultural Copout', *Time*, 26 January 2012 http://ideas.time.com/2012/01/26/nostalgia-our-favorite-cultural-copout/?iid=sr-link1 [accessed 6 November 2019].

United States Census Bureau, *State & County Quickfacts*, Michigan www.census.gov/quickfacts/table/PST045215/2622000 [accessed 20 October 2016].

Van Rensbergen, Henk, *Abandoned Places: The Photographer's Selection* (Tielt: Lannoo, 2014).

Vergara, Camilo José, *American Ruins* (New York: Monacelli Press, 1999).

Vergara, Camilo José, *The New American Ghetto* (New Brunswick, NJ: Rutgers University Press, 1995).

Vermeulen, Timotheus and Van Den Akker, Robin, 'What Meta Does and Does Not Mean', *Notes on Metamodernism*, 14 October 2010, www.metamodernism.com/2010/10/14/what-meta-means-and-does-not-mean/ [accessed 6 November 2019].

The Virgin Suicides, dir. by Sofia Coppola (Paramount Classics, 1999).

Wallace, David Foster, 'E Unibus Pluram: Television and U.S. Fiction', in *A Supposedly Fun Thing I'll Never Do Again: Essays and Arguments* (London: Abacus, 1998), pp. 21–82.

Wallerstein, Immanuel, *After Liberalism* (New York: New Press, 1995).

Wenders, Wim, *Places, Strange and Quiet* (Ostfildern: Hatje Cantz, 2013).

West, Nancy Martha, *Kodak and the Lens of Nostalgia* (Charlottesville: University of Virginia Press, 2000).

'The Wheel', *Mad Men*, dir. by Matthew Weiner (Debmar-Mercury, 2007).

Widdicombe, Lizzie, 'Tavi Says: Fashion Dictates from a Fourteen-Year-Old', *The New Yorker*, 20 September 2010, www.newyorker.com/magazine/2010/09/20/tavi-says [accessed 6 November 2019].

Wieseltier, Leon, 'Preface', in *Illuminations*, by Walter Benjamin (New York: Random House, 2002), pp. vii–x.

Williams, Raymond, 'Base and Superstructure in Marxist Cultural Theory', *New Left Review*, 1.82 (1973), 3–16.

Willis, Ellen, 'Three Elegies for Susan Sontag', in *The Essential Ellen Willis*, ed. by Nona Willis Aronowitz (Minneapolis: University of Minnesota Press, 2014), pp. 464–473.

Wiseman, Eva, 'Tavi Gevinson: The Fashion Blogger Becoming the Voice of a Generation', *Guardian*, 9 December 2012, www.theguardian.com/fashion/2012/dec/09/tavi-gevinson-fashion-blogger [accessed 6 November 2019].

Witt, Emily, 'Tavi Forever', *T Magazine: New York Times Style Magazine*, 6 June 2014, www.nytimes.com/2014/06/06/t-magazine/tavi-gevinson-on-rookie-magazine-and-growing-up.html?_r=0 [accessed 6 November 2019].

Woodward, Richard B. 'Disaster Photography: When is Documentary Exploitation', *Art News*, 2 June 2013, www.artnews.com/2013/02/06/the-debate-over-ruin-porn/ [accessed 6 November 2019].

Yeats, William Butler, 'He Wishes for the Cloths of Heaven', *Online-literature.com*, www.online-literature.com/frost/776/ [accessed 6 November 2019].

Žižek, Slavoj, 'Preface', in *The Privatization of Hope: Ernst Bloch and the Future of Utopia*, ed. by Peter Thompson and Slavoj Žižek (Durham, NC: Duke University Press), pp. xv–xx.

Žižek, Slavoj, 'You May!', *London Review of Books*, 18 March 1999, www.lrb.co.uk/v21/n06/slavoj-zizek/you-may [accessed 6 November 2019].

Index